FORMAL METHODS FOR
FOR
VLSI DESIGN

FORMAL METHODS FOR FOR VLSI DESIGN

IFIP WG 10.5 Lecture Notes

Edited by

JØRGEN STAUNSTRUP
Department of Computer Science
Technical University of Denmark
Lyngby, Denmark

1990

NORTH-HOLLAND
AMSTERDAM · NEW YORK · OXFORD · TOKYO

ELSEVIER SCIENCE PUBLISHERS B.V.
Sara Burgerhartstraat 25
P.O. Box 211, 1000 AE Amsterdam, The Netherlands

Distributors for the United States and Canada:
ELSEVIER SCIENCE PUBLISHING COMPANY INC.
655 Avenue of the Americas
New York, N.Y. 10010, U.S.A.

ISBN: 0 444 88858 6

Printed in The Netherlands.

Preface

These lecture notes have been written for the international summer school on "Formal Methods for VLSI Design" which took place in June 1990. The summer school was organized by the Computer Science Department at the Technical University of Denmark in cooperation with IFIP's working group 10.5 on VLSI.

The lecture notes contain an overview of an exciting new field of research. Formal methods aim at giving the VLSI designer a firm foundation and useful tools for developing integrated circuits. The methods provide the possibility of doing a systematic verification in the early phases of the design process. By verifying high level descriptions of the design before concerning himself with low level details, the designer can avoid wasting time implementing circuits that would later be discarded. This is advantageous, since it can be very expensive to locate and correct errors found in the later stages of a project; for example, if correcting these errors requires extensive, global changes to the design. Furthermore, the long turn-around time for circuit fabrication makes it attractive to use techniques which uncover errors at an early phase of the design.

The summer school consisted of six series of lectures each presenting a distinct formal method. Each is presented in a separate chapter of this book (one of Graham Birtwistle's lectures was given by Lars Rossen, this material is presented in a separate chapter). In addition to the lectures, there were hands on demonstrations of software tools supporting the formal methods.

It is a pleasure to acknowledge the support which was received from various sources. Without this, it would not have been possible to organize the summer school and publish this book.

- A generous grant by The Danish Research Academy made it affordable for Ph.D. students to participate in the summer school.

- Nokia Denmark provided Sun workstations used for demonstrations.

- Many members of IFIP WG 10.5 were very helpful in giving advice and distributing the information about the summer school.

Bente Falshøj and Marian Frandsen put a great effort into organizing the summer school.

Contents

Formal Methods for VLSI Design
J. Staunstrup (Editor)
Elsevier Science Publishers B.V. (North-Holland)
© IFIP, 1990

Chapter 1

Circuit design in Ruby

G. Jones[1] and M. Sheeran[2]

1.1 Introduction

The business of making integrated circuits is a peculiarly difficult engineering task largely because it spans such a wide range of levels of abstraction. Figure 1.1 suggests the sorts of names that might be used to describe these layers. A good design process must necessarily insulate each step of the design – as far as is possible – from the concerns that are most readily understood and tackled at other levels. As used here, the word *designing* simply means making progress down the spectrum from specification to implementation.

On the whole, we are not going to be concerned with details of implementation like gates and transistors. So to us the word *circuit* means no more than the sort of algorithm that might well be implemented by something like an integrated circuit or a collection of them. Designing circuits is the business of filling in the gap in the figure between specification capture and circuit fabrication.

That means, for example, that we will be interested in highly parallel algorithms with a more or less static structure; also that we are going to be concerned to minimize expensive sorts of communication, like broadcasting, and in emphasizing locality. We will not, however, be at all concerned with the details of any fabrication technology; nor in such concerns as how many transistors can be fabricated on a single chip.

---

[1] Oxford University Computing Laboratory, 11 Keble Road, Oxford OX1 3QD, England
[2] Department of Computing Science, University of Glasgow, Glasgow G12 8QQ, Scotland

Vague specification
Non-executable specification   ↓ capture
Abstract algorithm
Word-level algorithm
Bit-level algorithm
Gates
Transistors
Rectangles   fabrication
Silicon

Figure 1.1: the design process

It is quite common for the design process to be realized by a number of
steps passing from one level to a lower level, where the designer 'invents' the
implementation at the next level and then proves – after the fact – that this
new design does indeed implement the specification at the preceding level.
These *proof* steps would be upwards in the figure.

On the other hand, some steps of the design process consist of a purely
mechanical transformation of specification into implementation: for example,
the turning of a set of masks into a pattern on silicon. These can be thought
of as *compilation* steps, and require no ingenuity or invention. Indeed, they
are usually tedious and are best left to be done by a machine.

We tend to use of the word *calculation* to mean stages in the design process
which proceed forwards, without large inventive leaps and subsequent proofs
of correctness, but yet without necessarily being purely mechanical. Ruby is
a framework in which to do these calculations.

1.1.1   The shape of this chapter

This chapter introduces Ruby, and a calculational style in which to explore the
design of digital signal processing circuits and similar devices and algorithms.

The fundamental idea introduced by §1.2 is that circuits are built from
parts by a process of *composition*, and that this composition has simple mathematical properties like those of the composition of functions and the composition of relations. That gives us a way of describing circuits which are
composed 'sequentially' by being connected to each other, and in contrast
§1.3 describes a 'parallel' composition of parts which operate independently
on structured inputs: lists and tuples of signals.

The reality is that most circuit structures lie somewhere between these two
extremes, being only partly interconnected and partly independently parallel.
There are particular patterns of interconnection that arise often, and which
make sense under the constraints on the feasibility of layout. In the subsequent

sections such patterns of interconnection are discussed in Ruby, and we explore the mathematical properties of circuits which are composed in these ways. The idea is to become familiar with these properties so that they can be used to explore design choices in developing a circuit.

One of the important characteristics of Ruby is that it is no more difficult to deal with sequential circuits – ones with internal state – than it is to deal with combinational circuits. The design is done by the same sort of calculation in both cases, in the same sort of algebra. In §1.6 the interpretation of Ruby expressions as sequential circuits is introduced. We then go on to discuss manipulations like pipelining and data-skewing which are specific to sequential circuits.

By this point you should have enough Ruby to be able to tackle a sizeable example, and §1.7 does just that. It outlines a path through the design of a systolic correlator, from initial decisions about how to represent the interface to the consequences this has for the final circuit, and ultimately to optimizations of timing performance.

All of the circuit forms discussed up to that point are essentially iterative, but §1.8 shows that the same sorts of ideas are applicable to circuits like the 'butterfly' network that are naturally described in a recursive fashion. Butterfly networks are common in many sorts of circuits, but are perhaps most familiar from implementations of the 'fast Fourier transform'. Finally §1.9 shows that the butterfly implementation of the FFT can be discovered by calculation from the specification in just the same was as we did with the correlator.

1.1.2   The rôle of pictures

Throughout this chapter there are pictures that are intended to motivate the discussion that goes with them. At least for some people, the pictures help with an intuitive understanding of what an equation means, or of why a calculation proceeds in a certain way.

These 'abstract floorplans' share many of the important properties of the circuits with which a designer would be concerned. Principal amongst these is *locality*: whether or not components are connected to components that are near to them. Because the costs which dominate in circuit construction are those of communication, locality is very important. Similarly, the extent to which connections cross over each other shows up, and this is another cost in circuit construction.

Beware, however, of reasoning about the pictures. In particular, notice that quite often the pictures are just particular instances of the equations and calculation which they portray. There is a danger in trying to reason from the example in the picture that you might generalize a result which happens only to be true for the particular instance.

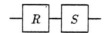

Figure 1.2: the composition $R\,;S$

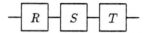

Figure 1.3: composition is associative: $(R\,;S)\,;T = R\,;S\,;T = R\,;(S\,;T)$

1.2 Composition and inverse

The central idea in Ruby is that of putting two circuits together to co-operate with each other. The composition of two circuits R and S will be written $R;S$ and you should have in mind the idea of a circuit in which connections are made from one side of R to the other side of S, rather as shown in figure 1.2. The 'left-hand' edge of a circuit will be called its *domain* and the 'right-hand' edge its *range*. The domain of $R\,;S$ is that of R, its range is that of S, and the range of R is connected to the domain of S. The fact that the domain appears on the left of the picture and the range on the right is of course just a convention, but will be one to which we adhere much of the time.

You should think of a circuit as being a relation which holds between the signals on its connections with the outside world. This is the relation which it enforces by arranging that the values of its outputs are consistent with those of its inputs. Beware the temptation to think of the domain signals as inputs and the range signals as outputs. The division between domain and range is a purely geographical one, and there may be inputs and outputs on either side of one of our circuits.

Composition of relations is defined by $x\,(R\,;S)\,z \iff \exists y.\,x\,R\,y\;\&\;y\,S\,z$. This is an associative operation, which is to say that the order in which the components of figure 1.3 are assembled cannot affect the meaning of the circuit. Associativity gives us an excuse for leaving out the parentheses in, for example, $R\,;S\,;T$.

Later when we come to deal with sequential circuits it will prove necessary to change our idea of what a signal is, and indeed of what a circuit is; but the things that stay the same are the laws about circuit forms. So, for example, composition will stay associative. It will pay to start as we mean to go on. You should try to avoid reasoning about the data – x, y, z in this example – and as far as possible, you should try to avoid reasoning about the specific component circuits – R, S, T here – concentrating instead on the combining forms, like composition, and the laws which they obey. The associative law – that $(R\,;S)\,;T = R\,;(S\,;T)$ for any R, S and T – is not about the data on which the circuit operates, nor is it about any particular circuit components R, S and T; rather it is a general statement about ';'.

1.2.1 Repeated composition

Where it makes sense to plug two components of the same kind together, we can talk about the n-times repeated composition of R, written R^n, and which is defined by $R^1 = R$, and $R^{n+1} = R$; R^n.

Since we know that composition is associative, we already know many things about repeated composition, for example $R^{n+1} = R^n$; R, and $R^{m+n} = R^m$; R^n, and so on. The proofs can be done by induction on n, and notice that the proof depends only on the associativity of composition, so there is no need to appeal to the meaning of R at all.

Suppose we assume for some R and S that R ; $S = S$; R, then a simple induction shows that R ; $S^n = S^n$; R. A similar induction shows that on the same assumption $(R ; S)^n = R^n$; S^n.

Now, notice that although R ; $S^n = S^n$; R and $(R ; S)^n = R^n$; S^n are only proved in case R and S commute (in case R ; $S = S$; R), and so are statements about R and S; nevertheless the theorems 'If R ; $S = S$; R then R ; $S^n = S^n$; R' and 'If R ; $S = S$; R then $(R ; S)^n = R^n$; S^n' are – just like the associative law – statements about composition.

These statements about composition can be added to the collection of laws that we are accumulating, and which will eventually constitute a body of knowledge about the *forms* of all circuits.

1.2.2 Inverse

The inverse of a circuit is its left-to-right reflection: the connections on the left of R appear on the right of R^{-1} and vice versa. Since reflecting twice gets us back the original circuit, we require that for any circuit R the inverse of its inverse should be the same circuit: $(R^{-1})^{-1} = R$. Inverse is often called converse, and is defined by $x R^{-1} y \iff y R x$.

Consider the relation which represents negation of Booleans: T *not* F and F *not* T. Its inverse not^{-1} is of course the same relation. That means that $not^{-1} = not$, so you should be aware that 'reflecting the picture' of *not* leaves it unchanged. In particular, there is no idea that the inputs and outputs of a gate have been exchanged: the sense in which $not^{-1} = not$ abstracts from such concerns as which signals are inputs and which are outputs.

The next question to ask is how inverse interacts with composition. As shown in figure 1.4 you can reflect the picture of a composition by swapping the components and reflecting each of them, so keeping corresponding edges connected. Stated as a Ruby law, we have that $(R ; S)^{-1} = (S^{-1})$; (R^{-1}) for any R and S.

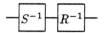

Figure 1.4: the inverse of a composition: $(R\,;S)^{-1} = (S^{-1})\,;(R^{-1})$

It follows that $(R^n)^{-1} = (R^{-1})^n$, so as a notational convenience we define $R^{-n} = (R^{-1})^n$. Be careful not to let this representational recklessness lead you astray – the induction that proves $R^{m+n} = R^m\,;R^n$ does so for positive n. Notice that since we know nothing in general about $R\,;R^{-1}$, it is not necessarily the case that $R^{m-n} = R^m\,;R^{-n}$. In this respect at least it can be argued that it was a mistake to christen the reflection of a circuit its 'inverse' with all the connotations that has.

1.2.3 Identity and types

Having defined R^n for positive and negative n, it is natural to seek an interpretation for R^0 which fits neatly into the remaining gap. The time has come to be honest about types. (Perhaps, since we are talking about 'circuits' and not relations, it might have been better to talk of '*interfaces*' rather than types.)

As far as will concern us, a type is just a collection of values. In the world of relations, the things that look like types are equivalence relations: an equivalence relation determines a set of equivalence classes, and the set of equivalence classes determines the equivalence relation. To describe what it is to be a type, we have to encode in Ruby the properties that make a relation be an equivalence relation.

An equivalence relation is one which is reflexive, symmetric and transitive: symmetry is just that $R = R^{-1}$; transitivity of a reflexive relation is just that $R = R^2$ (you can show that any transitive relation contains its square, but you need reflexivity to show containment of the relation in the square). So any R for which $R = R^{-1} = R^2$ (and so $R = R^n$ for all non-zero n) is an equivalence on all the relevant values; such an R we call a *type*. It will clearly do no harm to our intuition to define $R^0 = R$ so long as R is a type.

A type D is a type of the domain (not, notice, *the* type of the domain) of a circuit C if $C = D\,;C$, and similarly a type R is a type of its range if $C = C\,;R$. Sometimes this will be written as $C : D \to R$, and we speak of $D \to R$ as a type for C. When speaking of a circuit C, we ought really to quote a particular D and R, and speak of the circuit $C : D \to R$. Composition makes much more sense if you only ever compose circuits with matching types, say $P : A \to B$ and $Q : B \to C$, in which case $(P\,;Q) : A \to C$, as you can check from the definition.

In that case, it makes sense to talk of repeated composition only when you

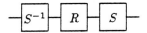

Figure 1.5: the conjugation $R \setminus S$ of R by S

have in mind a *homogeneous* component $R : T \to T$, in which case $R^n : T \to T$ for all positive and negative n. For that reason it is intuitively safe to define R^0 to be T, because then certainly $R^{n+0} = R^n = R^n ; T = R^n ; R^0$ and similarly for R^{0+n} and for negative powers. Be careful with this definition: strictly speaking it does not make sense to talk about R^0 without making explicit which particular type you have in mind for R – remember that the type is not unique – and a pedant would insist on saying that it is $(R : T \to T)^0$ that is defined to be T.

Of course the reason for being sloppy is the usual one: that it does not often matter all that much exactly what choice you make, and when it does the right choice is obvious. You should beware, however, of any intuition that leads you to expect R^0 necessarily to be the identity of composition. That would be an ι for which $\iota ; Q = Q = Q ; \iota$ for all Q.

1.2.4 Conjugation

A circuit and its inverse often appear together bracketing another component, as in figure 1.5. The conjugation of R by S is defined by $R \setminus S = S^{-1} ; R ; S$.

A conjugation like $R \setminus S$ behaves rather like the inside R part, and you can think of S as a 'translation' (in almost any of the many different senses of that word) applied to the language in which you communicate with R. For example, $(R \setminus S)^{-1} = R^{-1} \setminus S$, so the inverse of a conjugation of R is the same conjugation of the inverse of R.

Conjugations can be composed in the obvious way, because $(R \setminus S) \setminus T = R \setminus (S ; T)$. It is almost the case that the composition of two conjugations is the conjugation of the compositions. If $R ; R^{-1}$ is a type of the range of S and the domain of T, then

$$
\begin{aligned}
(S \setminus R) ; (T \setminus R) &= R^{-1} ; S ; (R ; R^{-1}) ; T ; R \\
&= R^{-1} ; (S ; T) ; R \\
&= (S ; T) \setminus R
\end{aligned}
$$

Such an R is called a *representation* of the type $R ; R^{-1}$, and R^{-1} is an *abstraction* to that type.

The idea behind these names is that we start from some operation defined on some abstract objects, say negation of Booleans: T *not* F and F *not* T. Suppose that we choose to represent Booleans by bits: each bit value is given

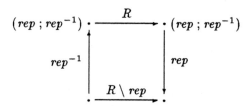

Figure 1.6: $R \setminus rep$ implements $R : (rep \,;\, rep^{-1}) \to (rep \,;\, rep^{-1})$

a meaning by the representation: say $\mathsf{F}\,rep\,0$ and $\mathsf{T}\,rep\,1$. You can calculate that $rep \,;\, rep^{-1}$ is the identity relation on the Booleans. Then $not \setminus rep$ is an implementation of the abstract negation operation, which if it is given a representation of a bit relates it to a representation of its negation. You can calculate that $0\,(not \setminus rep)\,1$ and $1\,(not \setminus rep)\,0$.

The condition that $rep \,;\, rep^{-1}$ be a type is just the requirement that if you take any abstract object of that type, then rep will give at least one concrete representative of that object, and that any representative given by rep will be a representative of that same abstract object. That is to say, rep gives a faithful representation of the type.

1.3   Lists and tuples

In order to talk about operations on collections of data, we need some structured types like lists and tuples. We are going to blur the distinction that people usually make between lists and tuples: there will be no difference between a triple of things that happen to have a common type, and a list of things that have that type and which happens to be three elements long. This makes some things easier, and some things – like mechanical type deduction – tremendously difficult.

1.3.1   Parallel composition

The fundamental operation that builds circuits that operate on lists is *parallel composition*. The parallel composition $[R, S]$ of R and S is a circuit whose domain and range are pairs of signals – as shown in figure 1.7 – which has an R component operating on the first elements of the pairs, and an S component independent of it operating on the second elements. That is to say $\langle x, y \rangle [R, S] \langle u, v \rangle$ exactly when both $x\,R\,u$ and $y\,S\,v$. Notice the following convention, to which the pictures in this chapter adhere: that the first component of a list of signals is on the left if you stand in the range of a circuit, looking towards

Figure 1.7: the parallel composition $[R, S]$

the domain. Since the range is on the right of figure 1.7, and the domain on the left, the first component is at the bottom and the last at the top.

The independence of the components of a parallel composition is reflected in the way that composition and inverse distribute over it, as $[R, S]$; $[T, U] = [R$; T, S ; $U]$ and $[R, S]^{-1} = [R^{-1}, S^{-1}]$. Beware the apparently misleading punctuation in $[R$; T, S ; $U]$, which can only mean $[(R$; $T), (S$; $U)]$.

1.3.2 Pairs and projections

The parallel composition of a circuit with an identity arises so often that we abbreviate the forms fst $R = [R, \iota]$, read 'first R', and snd $R = [\iota, R]$ read 'second R'. As you can calculate from the definition, a first and a second commute with each other, fst R ; snd $S =$ snd S ; fst R, and each distributes over composition and inverse, for $\mathsf{fst}(R$; $S) = $ fst R ; fst S and $(\mathsf{fst}\, R)^{-1} = $ fst R^{-1}, and similarly for second.

We will also often need to extract just one component of a pair, which we do with projections: π_1 projects a pair onto its first component, $\langle x, y \rangle\, \pi_1\, x$, and π_2 onto its second, $\langle x, y \rangle\, \pi_2\, y$. The inverse of a projection, of course, injects its domain into a pair (with an arbitrary other component) so that π_1^{-1} ; $\pi_1 = \iota$ is the universal type, as is π_2^{-1} ; π_2. The projections give us a way of talking about things that behave like pairs, and so of talking about operations on more than one argument. They satisfy $[R, S]$; $\pi_1 = $ snd S ; π_1 ; R and $[R, S]$; $\pi_2 = $ fst R ; π_2 ; S.

Beware that although fst R ; $\pi_1 = \pi_1$; R, in general $[R, S]$; $\pi_1 \neq \pi_1$; R for every S. If S is the empty relation then so is $[R, S]$, and so is any composition in which it appears, but π_1 ; R need not be empty. So it is the case that $(\mathsf{fst}\, R) \setminus \pi_1 = R$, but $[R, S] \setminus \pi_1 \neq R$.

1.3.3 Types for lists

Without going into any further formality, the notion of the parallel composition of two circuits extends naturally to the composition of any number of components. In particular, there is the parallel composition of one component alone, written $[R]$. The domain signal of this circuit is a singleton list, as is its range signal. Writing 1 for the type $[\iota]$ of singletons, $[R] : 1 \to 1$.

Figure 1.8: instances of *apl*, *app*, and *apr*$^{-1}$

In a picture it is hard to imagine there being any difference between an element of some type and a list of signals that consists of just one signal of that given type. The formalization of this similarity is that a list of one thing can be used to represent that thing, and any thing can be used to represent a list consisting of that thing alone. The abstraction that relates signals to singleton signals is here written $[-] : \iota \to 1$, which you can read as 'singleton' (or 'gift-wrap'). It is a representation for any type, which is to say that $[-] ; [-]^{-1} = \iota$, and it is an abstraction to singletons, which is to say that $[-]^{-1} ; [-] = 1$.

In order to talk about longer lists, more constants ('plumbing circuits') are needed. The thing that corresponds to appending lists is the *app* circuit for which

$$[R_0, R_1, \ldots, R_i, R_{i+1}, R_{i+2}, \ldots, R_n]$$
$$= \quad [[R_0, R_1, \ldots, R_i], [R_{i+1}, R_{i+2}, \ldots, R_n]] \setminus app$$

for any n and $i \leq n$. In terms of relations on signals, *app* relates a pair of lists to the list obtained by concatenating them; but note that all the names in this equation are names of circuits, rather than signals.

You may be more familiar with the idea of building up lists one component at a time: $apl = \mathsf{fst}\,[-] ; app$, read 'append left' or by some people 'cons', is a circuit which adds one signal to the left of a list; and $apr = \mathsf{snd}\,[-] ; app$, read 'append right' or 'snoc', adds a signal to the right-hand end of a list. For example, $[R, [S, T]] ; apl = apl ; [R, S, T]$ and $[[R, S], T] ; apr = apr ; [R, S, T]$ so $[R, [S, T]] ; apl ; apr^{-1} = apl ; apr^{-1} ; [[R, S], T]$. Notice that $[R, [S, T]]$ and $[[R, S], T]$ are different circuits, each relating pairs, and each is different from $[R, S, T]$ which relates triples.

Structures with one component have type $1 = [\iota]$. Those with $n + 1$ components have one more component than those of length n, and can be written as any one of $n+1 = \mathsf{snd}\,n \setminus apl = [1, n] \setminus app = [n, 1] \setminus app = \mathsf{fst}\,n \setminus apr$. Neither $apl ; apl^{-1}$ nor $apl^{-1} ; apl$ is the identity, but each is a type: $apl^{-1} ; apl$ is the type of lists of at least one element, and $apl ; apl^{-1}$ that of pairs with a list in the second element; similarly for *apr*. In order to complete the picture, the natural definition for $0 = [\,]$, the parallel composition of no component

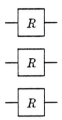

Figure 1.9: an instance of map R

circuits, makes it behave like the identity relation on zero length lists.

1.3.4 Map

The parallel compositions that can be written with the notation above are all circuits of a given, fixed width. So, for example, $[R, R]$ takes pairs of signals in the domain and in the range. The 'map' construction generalizes this, so that map R is the parallel composition of any number of R circuits. That means that 1 ; map $R = [R] = $ map R ; 1 and 2 ; map $R = [R, R] = $ map R ; 2, and so on. Figure 1.9 shows an instance of map R; this is an example of a picture from which it would be dangerous to generalize, because it has many properties not shared by, for example, (map $R) \setminus 137$, or (map $R) \setminus 1$.

By induction on the width you can show that map distributes over sequential composition and inverse in the same way that parallel composition does, so $\mathrm{map}(R \; ; \; S) = (\mathrm{map}\,R) \; ; \; (\mathrm{map}\,S)$ and $\mathrm{map}(R^{-1}) = (\mathrm{map}\,R)^{-1}$. It follows that if T is a type, then so is map T, in fact it is the type of all lists of signals of type T and in particular map ι is the type of lists.

There are a great number of things that you might want to be told about map to be sure that we were all thinking of the same circuit structure, amongst them perhaps

$$
\begin{aligned}
1 \; ; \mathrm{map}\,R &= [R] \\
n \; ; \mathrm{map}\,R &= \mathrm{map}\,R \; ; n \\
[\text{-}] \; ; \mathrm{map}\,R &= R \; ; [\text{-}] \\
app \; ; \mathrm{map}\,R &= \mathrm{map}\,\mathrm{map}\,R \; ; app \\
&= [\mathrm{map}\,R, \mathrm{map}\,R] \; ; app \\
apl \; ; \mathrm{map}\,R &= [R, \mathrm{map}\,R] \; ; apl \\
apr \; ; \mathrm{map}\,R &= [\mathrm{map}\,R, R] \; ; apr
\end{aligned}
$$

but it is clear that some of these are consequences of the others. In fact, you

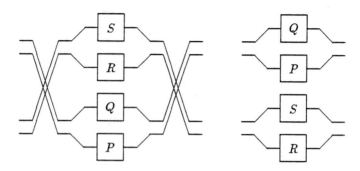

Figure 1.10: $[[P, Q], [R, S]] \setminus rev$ and the equivalent $[[R, S], [P, Q]]$

ought to be convinced by being told only that

$$
\begin{aligned}
{[\text{-}]} \,;\, \text{map}\,R &= R \,;\, [\text{-}] \\
[m, n] \,;\, app \,;\, \text{map}\,R &= [\text{map}\,R, \text{map}\,R] \,;\, [m, n] \,;\, app \\
(\text{map}\,R)^{-1} &= \text{map}(R^{-1})
\end{aligned}
$$

and perhaps some equations involving 0. These equations for all m and n are more than enough to define map.

1.3.5   Reverse

There is clearly a connection between left- and right-handed views of lists. Since any finite list could have been built up in either way, every time you say something about *apl* you are saying something about *apr*, and every time you say something about both there is a danger of saying something inconsistent about them.

To make it easier to talk about the mirror-similarity of *apl* and *apr*, we use a piece of plumbing *rev* : $\text{map}\,\iota \rightarrow \text{map}\,\iota$, read 'reverse', which flips a list over. In particular, $swap = rev \setminus 2$ exchanges two signals, so *rev* satisfies $rev\,;[R, S] = [S, R]\,;rev$. For other lengths of list it can be defined by $0\,;rev = 0$ and $[\text{-}]\,;\,rev = [\text{-}]$ and $app\,;\,rev = [rev, rev]\,;\,rev\,;\,app$.

Immediately from these definitions, you can calculate that $apl\,;\,rev = rev\,;\,\text{fst}\,rev\,;\,apr$ and $apr\,;\,rev = rev\,;\,\text{snd}\,rev\,;\,apl$. By induction on the length of the list, $rev = rev^{-1}$ and $rev\,;\,\text{map}\,R = \text{map}\,R\,;\,rev$ and so on.

Because *rev* reverses a list of signals, and because the convention about drawing circuits puts the beginning of a list at the bottom of the picture, the end of the list at the top, you might expect that a picture of $R \setminus rev$ would be the same as that of R but flipped over about a horizontal axis. This is almost right, but be careful: suppose R were a circuit operating on lists of lists, then the outermost list would be reversed but the elements in that list remain in

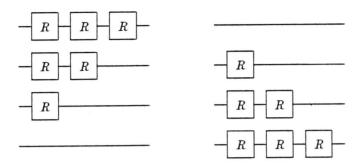

Figure 1.11: an instance of tri R, and an instance of irt R

the same order, so for example $[[P, Q], [R, S]] = [[R, S], [P, Q]] \setminus rev$ as can be seen in figure 1.10.

Reverse is the first example in Ruby of something that you might need to describe, but which it would be unpleasant to have to implement in a finished circuit. Any non-trivial instance of reverse drawn as a planar picture has to contain crossing wires: $rev \setminus 2$ contains exactly one crossing, $rev \setminus n$ contains $n(n-1)/2$. In the course of designing a circuit it will usually be an aim to eliminate instances of rev, and if that can not be done at least to reduce their width.

1.3.6 Triangle

A useful construction that is closely related to map, but which seems to be almost peculiar to the sorts of programs that arise in describing circuits, is the triangle. Instances of this are illustrated in figure 1.11. For any homogeneous $R : T \to T$, the triangle tri R : map $T \to$ map T relates lists of signals, just as map would, but it relates the i-th components according to R^i. For example, if $(\times 2)$ is the circuit which doubles a number, tri$(\times 2)$ relates a list of zeroes and ones to a list of those bits weighted by the powers of two which they would represent in a binary number with its least significant bit first. Similarly, $(\text{tri}(\times 2)) \setminus rev$ would give them the weights appropriate to having the most significant bit first. The form $(\text{tri } R) \setminus rev$ appears so often that we really need a name for it; in this chapter we adopt Robin Sharp's notation for it, which is irt R.

It should suffice to tell you that apl ; tri $R = [R^0, \text{map } R$; tri $R]$; apl, that $(\text{tri } R)^{-1} = \text{tri}(R^{-1})$, and that 0 ; tri $R = 0$; map R, but of course there is a rich collection of laws that can be derived from these and the properties of other operators by induction over lists. Among the more useful ones are those that relate triangles and maps of the same components such as $(\text{tri } R)^n = \text{tri}(R^n)$ and tri R ; map $R =$ map R ; tri R and those about commuting circuits, for

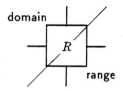

Figure 1.12: a different picture of $R : 2 \to 2$

if $R \,;\, S \,=\, S \,;\, R$ you can show that $\operatorname{tri} R \,;\, \operatorname{tri} S \,=\, \operatorname{tri}(R \,;\, S)$ and for types $\operatorname{tri}(R^0) = \operatorname{map}(R^0)$ and so $\operatorname{tri} R : \operatorname{map} R^0 \to \operatorname{map} R^0$.

Decomposing a triangle with *apr* instead of *apl*, or decomposing a reversed triangle with *apl*, you can show by induction on n that $\operatorname{fst} n \,;\, apr \,;\, \operatorname{tri} R = [n \,;\, \operatorname{tri} R, R^n] \,;\, apr$ and it turns out that this configuration appears often: an R^n will arise somewhere where by writing $(\operatorname{tri} R) \backslash apr^{-1}$ or $(\operatorname{irt} R) \backslash apl^{-1}$ you can avoid mentioning n altogether.

1.4   Rows and columns

There are fundamental reasons for being interested in circuits which can be laid out in a little more than two dimensions. Any very large system with many connections has to be more or less flat. This means that there are good reasons to be interested in two-dimensional meshes. This section is almost entirely about circuits $R : 2 \to 2$ with a pair of signals in the domain and a pair of signals in the range. To make pictures of networks of these circuits lie naturally on the page, they are drawn by following a new convention illustrated in figure 1.12. The first component of the domain lies to the left, the second component above the circuit; the first component of the range lies below the circuit, and the second component to its right. This keeps the order of the signals on the paper consistent with the convention given earlier for drawing circuits that operate on pairs.

1.4.1   Beside and below

There are two ways of connecting these square tiles that seem natural: as shown in figure 1.13. In each case the wires are grouped so that both $R \leftrightarrow S$, read 'R beside S', and $R \updownarrow S$, read 'R below S', are also of type $2 \to 2$; indeed $R \leftrightarrow S : \operatorname{snd} 2 \to \operatorname{fst} 2$ and $R \updownarrow S : \operatorname{fst} 2 \to \operatorname{snd} 2$.

It is clear that circuits made by beside and below are not new ones, in the sense that we already know enough wiring to be able to duplicate their function without any new primitives. Reading off the picture in figure 1.14,

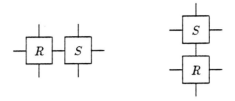

Figure 1.13: R beside S making $R \leftrightarrow S$, and R below S making $R \updownarrow S$

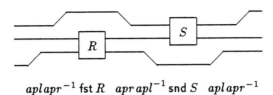

$apl\,apr^{-1}\,\mathsf{fst}\,R \quad apr\,apl^{-1}\,\mathsf{snd}\,S \quad apl\,apr^{-1}$

Figure 1.14: another circuit layout which implements $R \leftrightarrow S$

the definitions of beside and below would appear to be straightforward

$$R \leftrightarrow S = apl\,;((\mathsf{fst}\,R)\setminus apr)\,;((\mathsf{snd}\,S)\setminus apl)\,;apr^{-1}$$
$$R \updownarrow S = (R^{-1} \leftrightarrow S^{-1})^{-1}$$
$$= apr\,;((\mathsf{snd}\,S)\setminus apl)\,;((\mathsf{fst}\,R)\setminus apr)\,;apl^{-1}$$

Unfortunately, these definitions would not be very useful: this is an example of recklessly generalizing from a picture and getting more than one bargains for.

One of the properties we expect of beside and below is that $(A \leftrightarrow B) \updownarrow (C \leftrightarrow D) = (A \updownarrow C) \leftrightarrow (B \updownarrow D)$. Richard Bird [Bird88] describes this property by saying that 'beside *abides with* below', a contraction of above-and-besides. It turns out that the operators defined above do not abide: the abides property of beside and below depends on the component circuits being of type $2 \to 2$. Better definitions would therefore be given by

$$R \leftrightarrow S = apl\,;((\mathsf{fst}(R \setminus 2))\setminus apr)\,;((\mathsf{snd}(S \setminus 2))\setminus apl)\,;apr^{-1}$$
$$R \updownarrow S = apr\,;((\mathsf{snd}(S \setminus 2))\setminus apl)\,;((\mathsf{fst}(R \setminus 2))\setminus apr)\,;apl^{-1}$$

which is indeed an abiding pair of operators.

Particularly useful in discussing pair-to-pair circuits will be the above and beside of two identity components. Let $rsh = \iota \leftrightarrow \iota$, for 'right shift', and $lsh = \iota \updownarrow \iota$, for 'left shift'; clearly $rsh = lsh^{-1}$ and $lsh = rsh^{-1}$. Since $\iota \leftrightarrow \iota = apl\,;3\,;apr^{-1}$ you can see that $[a,[b,c]]\,;rsh = rsh\,;[[a,b],c]$, whence the name; and dually $lsh = apr\,;3\,;apl^{-1}$ and $[[a,b],c]\,;lsh = lsh\,;[a,[b,c]]$.

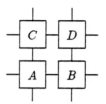

Figure 1.15: $(A \leftrightarrow B) \updownarrow (C \leftrightarrow D) = (A \updownarrow C) \leftrightarrow (B \updownarrow D)$

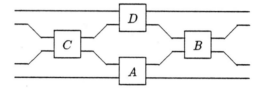

Figure 1.16: $([\iota, C, \iota] \setminus mid^{-1}) ; [A, D] ; ([\iota, B, \iota] \setminus mid^{-1})$

The definitions of beside and below can be recast in terms of left and right shifts, for $R \leftrightarrow S = rsh ; \text{fst } R ; lsh ; \text{snd } S ; rsh$ and by duality $R \updownarrow S = lsh ; \text{snd } S ; rsh ; \text{fst } R ; lsh$.

The easiest proof that beside abides with below goes by showing that each of $(A \leftrightarrow B) \updownarrow (C \leftrightarrow D)$ and $(A \updownarrow C) \leftrightarrow (B \updownarrow D)$ is equal to the same symmetrical form; for

$$(A \leftrightarrow B) \updownarrow (C \leftrightarrow D) = ([\iota, C, \iota] \setminus mid^{-1}) ; [A, D] ; ([\iota, B, \iota] \setminus mid^{-1})$$

where mid is the unique $mid : [2, 2] \to [\iota, 2, \iota]$ for which $[[a, b], [c, d]] ; mid = mid ; [a, [b, c], d]$. A circuit layout suggestive of this symmetrical form is shown in figure 1.16. You can then show by taking inverses throughout that

$$(A \updownarrow C) \leftrightarrow (B \updownarrow D) = ([\iota, C, \iota] \setminus mid^{-1}) ; [A, D] ; ([\iota, B, \iota] \setminus mid^{-1})$$

which shows that beside abides with below.

The other things that one may want to know about beside and below also say that it does not matter how you choose to bracket a picture, such as those in figure 1.17. First it is the case that $(R ; \text{snd } X) \leftrightarrow S = R \leftrightarrow (\text{fst } X ; S)$ and similarly by inverting both sides $R \updownarrow (S ; \text{fst } X) = (\text{snd } X ; R) \updownarrow S$. In particular, if $P ; Q = Y ; Z$ and $R = R ; \text{snd } Y$ and $\text{fst } Z ; S = S$ then $(R ; \text{snd } P) \leftrightarrow (\text{fst } Q ; S) = R \leftrightarrow S$.

Secondly it is the case that $[A, [B, C]] ; (R \leftrightarrow S) ; [[D, E], F] = ([A, B] ; R ; \text{fst } D) \leftrightarrow (\text{snd } C ; S ; [E, F])$ and dually $[[A, B], C] ; (R \updownarrow S) ; [D, [E, F]] = (\text{fst } A ; R ; [D, E]) \updownarrow ([B, C] ; S ; \text{snd } C)$. Most of the things that one needs to prove about these circuits are consequences of these last observations, setting various of the variables to the identity.

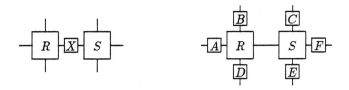

Figure 1.17: $(R \; ; \mathsf{snd}\, X) \leftrightarrow S = R \leftrightarrow (\mathsf{fst}\, X \; ; S)$ and $[A, [B, C]] \; ; (R \leftrightarrow S) \; ;$
$[[D, E], F] = ([A, B] \; ; R \; ; \mathsf{fst}\, D) \leftrightarrow (\mathsf{snd}\, C \; ; S \; ; [E, F])$

1.4.2 Reflections

Throughout the discussion of 'beside' and 'below' we made use of the duality
of the two operators. Taking the inverse of a $2 \to 2$ circuit corresponds to
flipping the picture about the diagonal line between the domain and range
– the line shown in figure 1.12. Since this turns vertical domain connections
into horizontal range connections and vice versa, it turns below into beside
and vice versa. That is the duality expressed by $(R \leftrightarrow S)^{-1} = (R^{-1}) \updownarrow (S^{-1})$.

Often, we find ourselves wanting to swap two of the wires of one of these
four-sided tiles. Define $\langle a, b \rangle \, \mathsf{swp}\, R \, \langle c, d \rangle = \langle d, b \rangle \, R \, \langle c, a \rangle$, so that a picture of
$\mathsf{swp}\, R$ is the same as one of R except that the connections on the right and
left are exchanged. The connections at the top and bottom are exactly the
same. Amongst the useful things to know about swp are that

$$
\begin{aligned}
\mathsf{swp}\,\mathsf{swp}\, R &= R \setminus 2 \\
(\mathsf{swp}\, R) \leftrightarrow (\mathsf{swp}\, S) &= \mathsf{snd}\, swap \; ; \mathsf{swp}(S \leftrightarrow R) \; ; \mathsf{fst}\, swap \\
(\mathsf{swp}\, R) \updownarrow (\mathsf{swp}\, S) &= \mathsf{swp}(R \updownarrow S)
\end{aligned}
$$

Of course there is a dual operation: the circuit $(\mathsf{swp}\, R^{-1})^{-1}$ is the one that
you get by reflecting R about a horizontal line, with the top and bottom
connections exchanged and the horizontal ones unchanged. This operator
has similar properties, as you can check by taking inverses throughout the
equations describing swp.

1.4.3 Other orthogonally connected circuits

There are a number of other ways that circuits might be connected with
the wires lying orthogonally to each other, and that might be interesting: for
example it will be common to connect circuits with two inputs and one output
in rows when doing things that are like adding up a row of numbers. There
seem to be a vast number of possibilities corresponding to the combinations
of connections that might be missing from the picture of beside and below.
Three of these configurations are illustrated in figure 1.18.

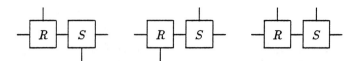

Figure 1.18: representation of $(R;S)\backslash 2$ and snd $\pi_2{}^{-1}$;$((\pi_1;R)\leftrightarrow(S;\pi_2{}^{-1}))$;fst π_1
and $((R\,;\,\pi_2{}^{-1})\leftrightarrow(S\,;\,\pi_2{}^{-1}))\,;\,\pi_2$

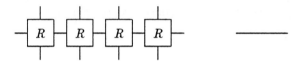

Figure 1.19: two instances of row R

Fortunately, however, these are all just instances of beside, for example
$(R\,;\,S)\setminus 2 =$ snd $\pi_1{}^{-1}\,;\,((R\,;\,\pi_2{}^{-1})\leftrightarrow(\pi_1\,;\,S))\,;$ fst π_2. This means that the
existing results about beside carry over immediately to any new operator of
this form that we might want. For example, suppose that $R\oplus S =$ snd $\pi_2{}^{-1}\,;$
$((\pi_1\,;\,R)\leftrightarrow(S\,;\,\pi_2{}^{-1}))\,;$ fst π_1, which is the first configuration shown in the
figure, then immediately from the properties of beside it follows that $[A,B]\,;$
$(R\oplus S)\,;\,[C,D] = (A\,;\,R\,;\,\mathsf{fst}\,C)\oplus(\mathsf{snd}\,B\,;\,S\,;\,D)$. It is the existence of the
inverses of the projections that make this unification possible, and this is one
of the excuses for not confining ourselves to functions, nor indeed to total
functions.

1.4.4   Rows

Having put two circuits next to each other with beside, the next step is of
course to put another next to them and so on, as shown in figure 1.19. This
is the same generalization that takes you from parallel composition to map,
and the following requirements on row should look very like those for map

$$\mathsf{snd}\,[\text{-}]\,;\,\mathsf{row}\,R \;=\; R\,;\,\mathsf{fst}\,[\text{-}]$$
$$\mathsf{row}\,R\,;\,\mathsf{fst}\,[\text{-}]^{-1} \;=\; \mathsf{snd}\,[\text{-}]^{-1}\,;\,R$$
$$\mathsf{snd}([m,n]\,;\,app)\,;\,\mathsf{row}\,R \;=\; ((\mathsf{row}\,R\,;\,\mathsf{fst}\,m)\leftrightarrow(\mathsf{row}\,R\,;\,\mathsf{fst}\,n))\,;\,\mathsf{fst}\,app$$
$$\mathsf{row}\,R\,;\,\mathsf{fst}(app^{-1}\,;\,[m,n]) \;=\; \mathsf{snd}\,app^{-1}\,;\,((\mathsf{snd}\,m\,;\,\mathsf{row}\,R)\leftrightarrow(\mathsf{snd}\,n\,;\,\mathsf{row}\,R))$$

Notice that these four conditions cannot be abbreviated in the same way as
those for map, by using inverse, because $(\mathsf{row}\,R)^{-1}\neq\mathsf{row}\,R^{-1}$.

The interface rule for row, by analogy with that for beside, would seem to
be that if $D\,;\,A = Y\,;\,Z$ where $R:\mathsf{fst}\,Z\to\mathsf{snd}\,Y$, then

$$[A,\mathsf{map}\,B]\,;\,\mathsf{row}\,R\,;\,[\mathsf{map}\,C,D] = \mathsf{row}([A,B]\,;\,R\,;\,[C,D])$$

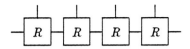

Figure 1.20: an instance of rdl R

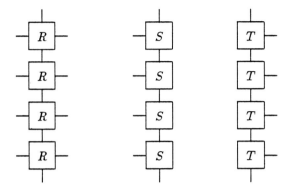

Figure 1.21: an instance of col R, one of rdr S, and one of $(\text{rdl}(T^{-1}))^{-1}$

By induction from the rule about a component lying between two circuits beside each other, it follows that fst X ; row$(R$; snd $X) = $ row(fst X ; $R)$; snd X. Similarly by induction from the rule relating swp and beside, we get a rule about reflecting a row, that swp row $R = $ snd *rev* ; row swp R ; fst *rev*.

There is a row-like repeated form that deserves a special mention, namely rdl R read 'reduce (from the) left'. This is illustrated in figure 1.20 and defined by rdl $R = $ row$(R;\pi_2^{-1});\pi_2$. Its properties are immediate from the corresponding properties of row. It should be suggesting operations like summation to you: for example, if $\langle x, y\rangle\, add\,(x+y)$ then $\langle a, \langle p, q, r\rangle\rangle\,(\text{rdl}\, add)(((a+p)+q)+r)$. The order of the bracketing – from the left – is important for non-associative operators, for example rdl *apr* or rdl(*swap* ; *apl*), and explains the name.

1.4.5   Columns

Of course since beside has below as its dual there is going to be a corresponding dual for row which is column, defined by col $R = $ (row $R^{-1})^{-1}$. An instance of this is illustrated in figure 1.21, and its properties are the duals of the properties of row.

$$\text{fst}\,[\text{-}]\,;\text{col}\,R = R\,;\text{snd}\,[\text{-}]$$
$$\text{col}\,R\,;\text{snd}\,[\text{-}]^{-1} = \text{fst}\,[\text{-}]^{-1}\,;R$$
$$\text{fst}([m, n]\,;\, app)\,;\text{col}\,R = ((\text{col}\,R\,;\text{snd}\,m)\,\updownarrow\,(\text{col}\,R\,;\text{snd}\,n))\,;\text{snd}\, app$$

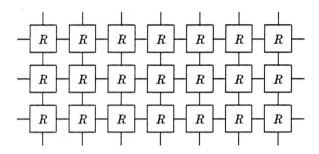

Figure 1.22: an instance of row col R = col row R

$$\text{col } R \, ; \text{snd}(\text{app}^{-1} \, ; [m, n]) \quad = \quad \text{fst } \text{app}^{-1} \, ; ((\text{fst } m \, ; \text{col } R) \updownarrow (\text{fst } n \, ; \text{col } R))$$

and so on.

The column-shaped thing that corresponds to reduce from the left is reduce from the right, defined by rdr R = col$(R \, ; {\pi_1}^{-1}) \, ; \pi_1$. It might seem that it would have been more logical to christen this something like 'reduce downwards', but again its name derives from the bracketing in the formula which describes the reduction of a function; for example, if $\langle x, y \rangle$ *add* $(x + y)$ then $\langle \langle p, q, r \rangle, a \rangle$ (rdr *add*) $(p + (q + (r + a)))$.

Beware that reduce right is not the inverse-dual of reduce left! The easiest way to see the difference is to compare the examples of rdr R and $(\text{rdl}(R^{-1}))^{-1}$ that are shown in figure 1.21. Of course $(\text{rdr}(R^{-1}))^{-1}$ is yet a fourth different reduce-shaped circuit.

Since for any $R : 2 \rightarrow 2$ both row $R : 2 \rightarrow 2$ and col $R : 2 \rightarrow 2$, it is possible to make row col R and col row R. Both of these will look like the example shown in figure 1.22, and in fact you can show by induction from the abides property of below and beside that row col R = col row R.

1.4.6 Horner's rule

Horner's rule is the name usually given to a method of evaluating polynomials without needing to raise powers or do unnecessary multiplications; it is encapsulated in the equality of $a_0 + x a_1 + \cdots + x^{n-2} a_{n-2} + x^{n-1} a_{n-1} + x^n a_n = a_0 + x(a_1 + \cdots + x(a_{n-2} + x(a_{n-1} + x a_n)) \cdots)$. Naïve evaluation of the left-hand side would appear to require a quadratic number of multiplications, that of the right-hand side only a linear number. The equality is a consequence of a property that is usually expressed as the distribution of multiplication over addition. Specifically it follows from $x \times a + x \times b = x \times (a + b)$.

The rule applies to any structure with two operations that distribute in this way, and in particular it has a counterpart in circuits:

$$[R, R] \, ; S = S \, ; R \quad \Longrightarrow \quad ((\text{tri } R) \setminus \text{apr}^{-1}) \, ; \text{rdr } S = \text{rdr}(\text{tri } R \, ; S)$$

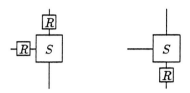

Figure 1.23: the hypothesis for Horner's rule, that $[R, R] \, ; S = S \, ; R$

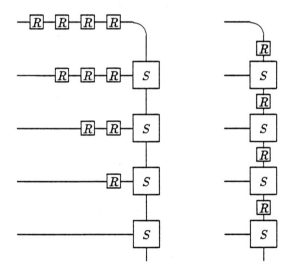

Figure 1.24: an instance of the consequence of Horner's rule, that tri $R; apr^{-1} \, ;$ rdr $S = apr^{-1} \, ;$ rdr(tri $R \, ; S)$

The hypothesis is illustrated in figure 1.23 and an example of the consequence for a particular size appears in figure 1.24.

The most familiar instance of Horner's rule will convert a list of bits into the number that they represent in binary notation. Suppose R is the circuit $(\times 2)$ that doubles a number, then recall that $\text{tri}(\times 2)$ relates a list of zeroes and ones to a list of those bits weighted by the powers of two which they would represent in a binary number with its least significant bit first. Suppose that S is the circuit *add* that relates a pair of numbers to their sum, then apr^{-1}; rdr *add* relates a list of numbers to their sum; so $bin = \text{tri}(\times 2)$; apr^{-1}; rdr *add* relates a list of bits to the number that they represent as a binary number, least significant bit first. (It also relates other lists of numbers to the same value, but that does not matter for the moment.)

The hypothesis of Horner's rule is satisfied by these two circuits, because the circuit on one side is $[(\times 2), (\times 2)]$; *add* which relates a pair of numbers to the sum of twice each of them, and that on the other is $\text{fst}(\times 2)^0$; add; $(\times 2) = add$; $(\times 2)$ which relates a pair of numbers to twice their sum. These are the same circuit because $2x + 2y = 2(x + y)$, so $[(\times 2), (\times 2)]$; $add = \text{fst}(\times 2)^0$; add; $(\times 2)$. Our formulation of Horner's rule therefore allows the conclusion that $bin = apr^{-1}$; $\text{rdr}(\text{tri}(\times 2)$; $add) = apr^{-1}$; rdr *step* where $step = \text{snd}(\times 2)$; add, which is to say that a bit-vector can be converted into the number that it represents by a right-reduction of *step* components, each of which doubles one input and adds it to the other.

There is a statement of Horner's rule for left reduction:

$$[R, R]\, ;\, S = S\, ;\, R \quad \Longrightarrow \quad ((\text{irt}\, R) \setminus apl^{-1})\, ;\, \text{rdl}\, S = \text{rdl}(\text{irt}\, R\, ;\, S)$$

Similarly there are statements for the inverses of both reductions. There are also a variety of similar results for row and column, and for grids. Their proofs by induction would be tedious and lengthy, but the whole point of this work is that such proofs once having been done, their consequences can be applied in the course of a design without the necessity of recourse to induction.

1.5   Transposition and zips

So far we have only seen wires crossed over in *rev*, but there is another cliché of circuit design which is even worse from the point of view of crossed wires. Suppose a circuit takes its input from a pair of buses, and calculates a function of the signals on pairs of wires drawn from corresponding positions in those buses. Somewhere in that circuit the designer has to arrange to interleave the buses, along the lines shown in figure 1.25, and that costs a number of wire crossings quadratic in the widths of the buses.

To describe such circuits in Ruby we need a new piece of plumbing, *zip*, defined by $\langle x, y \rangle zip\, z \iff \forall i.\, z_i = \langle x_i, y_i \rangle$. The law about circuit constructors

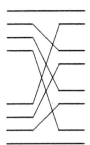

Figure 1.25: an instance of *zip*

that corresponds to this is that $[\text{map } R, \text{map } S]\,;zip = zip\,;\text{map}[R, S]$. Another way of saying this is that $[\text{map } R, \text{map } S] \setminus zip = \text{map}[R, S]$, but be careful because $(\text{map}[R, S]) \setminus zip^{-1} \neq [\text{map } R, \text{map } S]$. Similarly $[\text{tri } R, \text{tri } S] \setminus zip = \text{tri}[R, S]$, but not the other equation. This is because you can only interleave buses that are the same width, that is $zip^{-1}\,;zip = \text{map } 2$ but $zip\,;zip^{-1}$ is the type of a pair of lists that have the same length.

More generally any number of buses can be interleaved by transposing them with *trn*, defined by $x \; trn \; y \iff \forall i, j.\, x_{i,j} = y_{j,i}$, so $trn = trn^{-1}$, but be careful because $trn\,;trn^{-1} \neq \text{map map } \iota$. Since $zip = 2\,;trn = trn\,;\text{map } 2$ there are laws like those about zipping but about transposition and map, such as $\text{map map } R\,;trn = trn\,;\text{map map } R$ and more suggestively $\text{map tri } R\,;trn = trn\,;\text{tri map } R$ and so on.

Of course because zip and transpose are 'expensive' it is a good idea to avoid them in the final form of a circuit, but having them in the language makes it much easier to do some calculations. One has to be able to say what it is that one would not want to build, as well as what to aim for.

1.5.1 Zipping rows together

One of the common optimizations in circuit design is to interleave a number of similar regular grid-shaped calculations, such as arise in dealing with binary representations of numbers, to shorten the wires that connect them, for example bringing bits of the same weight in different numbers together if they are to be added. This is the transformation that takes a number of arithmetic circuits and interleaves them to make the data-path of a processor.

In Ruby the validity of this transformation is captured by the observation that $([A, B] \setminus zip^{-1}) \leftrightarrow ([C, D] \setminus zip^{-1}) = \text{snd } zip^{-1}\,;([A \leftrightarrow C, B \leftrightarrow D] \setminus zip^{-1})\,;$ $\text{fst } zip$ and of course the dual result for below. It generalizes by induction to

$$\text{row}([R, S] \setminus zip^{-1}) = \text{snd } zip^{-1}\,;([\text{row } R, \text{row } S] \setminus zip^{-1})\,;\text{fst } zip$$
$$\text{col}([R, S] \setminus zip^{-1}) = \text{fst } zip^{-1}\,;([\text{col } R, \text{col } S] \setminus zip^{-1})\,;\text{snd } zip$$

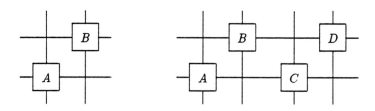

Figure 1.26: $[A, B] \setminus zip^{-1}$ and $([A, B] \setminus zip^{-1}) \leftrightarrow ([C, D] \setminus zip^{-1})$

$$\mathsf{row}((\mathsf{map}\ R) \setminus zip^{-1}) = \mathsf{snd}\ trn^{-1}\ ;\ ((\mathsf{map}\ \mathsf{row}\ R) \setminus zip^{-1})\ ;\mathsf{fst}\ trn$$
$$\mathsf{col}((\mathsf{map}\ R) \setminus zip^{-1}) = \mathsf{fst}\ trn^{-1}\ ;\ ((\mathsf{map}\ \mathsf{col}\ R) \setminus zip^{-1})\ ;\mathsf{snd}\ trn$$

and by application of two of the above to such observations as

$$\mathsf{row}\ \mathsf{col}((\mathsf{map}\ R) \setminus zip^{-1}) = ((\mathsf{map}\ \mathsf{row}\ \mathsf{col}\ R) \setminus zip^{-1}) \setminus [trn, trn]$$

1.6   Sequential circuits

So far we have intended to give the impression that all the circuits that we are talking about are combinational: that they behave as described in a steady state, in which they are presented with an unchanging input and deliver an unchanging output which depends only on that input. Of course no circuit is ever used in this way: at the very least there was a time before it was presented with any input; however the combinational description faithfully models the behaviour of a circuit that was presented with its input so long ago that it has settled down.

If an entire system is allowed to settle after each change in its input, the whole of the system is tied up for a settling time that depends on the longest propagation path, so it usual to place 'latches' – sometimes called 'registers' – at the outputs of the modules that make up a system. Each latch stores the output from its module in the previous steady state of that module, allowing that module to proceed with the next calculation while the modules that follow it are still settling into the previous steady state. In figure 1.27, suppose that the F is presented with a succession x_i of inputs. The output from the latch $\mathcal{D}$ at any time will be the value that was previously presented to it by F, so G can be completing the calculation of $G(F(x_i))$ while F is working on the calculation of $F(x_{i+1})$. When the circuit has settled the latch is 'clocked' – it is told to discard its current output and to propagate $F(x_{i+1})$ instead – by a signal not shown in the diagram. This technique of overlapping a succession of combinational calculations is called 'pipelining'. Since it has the effect of shortening the longest time to settle, it allows a greater number of calculations to be completed by a circuit in a given time. Notice, however,

Figure 1.27: a pipeline F ; $\mathcal{D}$; G

that there is no longer a single combinational description of what the whole circuit is doing.

It is usual to think of the latches in such a circuit as being the repository of the state of the circuit. The calculation being implemented by the circuit at any given time depends not only on the circuit and the inputs to it, but also on the values being held in latches. By means of these the calculation at any given time can be influenced by past calculations. In a pipeline like that in figure 1.27 the calculation, and so the output, depends only on a finite number of past inputs, in this case on only one; but in general there may be cycles in the circuit with the input to a latch depending on some function of its output. In that case the state held in the latch is an accumulation of information about the whole past history of calculations performed by the circuit, and the output can depend on an arbitrarily large number of past inputs. There is no way of giving a convenient and convincing account of such circuits in terms of a succession of independent combinational calculations.

The process of designing a sequential circuit appears to be significantly harder than that of designing a combinational one: even in the simple case of a pipeline there is an additional concern, that of making sure that the inputs that arrive together at a component are those for the same calculation; and more generally it is not possible to separate the concern of getting the time right from that of getting the value right. Fortunately, there is a large range of circuits for which it is possible to reason that a design which would work as a combinational circuit would work more or less the same way as a sequential circuit.

1.6.1 Time sequences

The sequential circuits with which we are concerned are those about which it is sound to reason in exactly the same way as those in previous sections. What that means is that the algebra is going to be the same as it was in previous sections, but the interpretation of the symbols will have to be different: the variables like R, S and so on will represent sequential circuits; and the combining forms will be correspondingly different.

Think of a circuit as relating not values, but time-sequences of values. A time sequence is just a function from an index which represents the time at which a signal is observed to the value that the signal has at that time. We will use the integers to represent successive times at which a signal is changed:

a signal representing a Boolean, one of type *bool*, is a function $Z \to B$, and one representing a pair of Booleans, one of type $[bool, bool]$, is a function $Z \to (B \times B)$, that is a pair-valued function, and *not* a pair of functions. So for example to say that *not* : *bool* $\to$ *bool* is to say that in this model it represents a relation between a function of type $Z \to B$ and another of the same type, and to say that *xor* : $[bool, bool] \to bool$ is to say that it represents a relation between a function of type $Z \to (B \times B)$ and one of type $Z \to B$.

Which relation should *xor* be in this model? It should relate a pair of Booleans to their logical 'exclusive or', and it should do this at each time. Similarly, a sequential *not* should, at each instant, relate a Boolean to its negation.

Let $\mathcal{S}[\![R]\!]$ be the relation that R represents in the sequential model, and $\mathcal{C}[\![R]\!]$ be the relation that it represented in the combinational model, then provided R has no internal state, we want $x \, \mathcal{S}[\![R]\!] \, y \iff \forall i. \, x_i \, \mathcal{C}[\![R]\!] \, y_i$, and we say that any R for which this is true is *stateless*.

Note that in the sequential model a circuit has a single time sequence in each of its domain and range. So, for example, a sequential *xor* relates a sequence of pairs of Booleans to a sequence of Booleans.

1.6.2   Composition, parallel composition and so on

Having decided what the interpretation of some circuits will be in the sequential model, what should the interpretation of composition be? At least for stateless circuits it is already determined, for if R and S are stateless we certainly want $R \, ; S$ to be stateless, so the only possibility is to define $\mathcal{S}[\![R \, ; S]\!]$ to be $\mathcal{S}[\![R]\!] \, ; \mathcal{S}[\![S]\!]$ for general R and S. Similarly, the inverse of a sequential circuit is interpreted as the relational converse of its interpretation: $\mathcal{S}[\![R^{-1}]\!] = \mathcal{S}[\![R]\!]^{-1}$, and so also for repeated composition.

In order to talk about the interpretation of parallel composition, we need to be able to relate a pair of sequences to the obvious sequence of pairs to which it corresponds: we already have a suitable relation, *zip* defined by $\langle x, y \rangle \, zip \, z \iff \forall i. \, \langle x_i, y_i \rangle = z_i$ and we define $\mathcal{S}[\![[R, S]]\!] = [\mathcal{S}[\![R]\!], \mathcal{S}[\![S]\!]] \setminus zip$ for general R and S. In giving an interpretation for map, we use *trn* to convert a time-sequence of n-sequences into an n-sequence of time-sequences. and define $\mathcal{S}[\![map \, R]\!] = (map \, \mathcal{S}[\![R]\!]) \setminus trn$ for general R. The interpretation of the other combining forms follows from these.

1.6.3   Delay and state

So far all the sequential circuits we can talk about have been stateless because all our combining forms preserve statelessness. In order to talk about circuits with state it will be enough for the present to introduce one new primitive: a *delay*, written $\mathcal{D}$. Its meaning in the sequential model is the relation

$$x \, \mathcal{D} \, y \iff \forall i. \, x_i = y_{i+1}$$

which says that delay relates two signals if the range signal is at all times the same as the domain signal had been exactly one time step earlier.

The inverse of a delay, an *anti-delay* written of course $\mathcal{D}^{-1}$, is also a useful concept. It represents the relation $x \, \mathcal{D}^{-1} \, y \iff \forall i. \, x_i = y_{i-1}$ which says that it relates two signals if the range signal is at all times the same as the domain signal would be exactly one time step later.

If these relations are to be implemented by circuits taking inputs from their domains and returning outputs in their ranges, then $\mathcal{D}$ is just a latch: the output after each clock is the same as the input immediately preceding it. The corresponding interpretation of $\mathcal{D}^{-1}$ as a circuit is altogether less implementable: it would have to predict *before* each clock what its input was going to be *after* that clock.

Conversely, if the relations are to be implemented by circuits taking inputs from their ranges and returning outputs in their domains, $\mathcal{D}^{-1}$ is a latch, and it is $\mathcal{D}$ that becomes unimplementable. What this means is that if an expression including $\mathcal{D}$ or $\mathcal{D}^{-1}$ were to be implemented by translating it into a circuit there would be these additional constraints on the choice of direction for each of the signal flows, that all $\mathcal{D}$ and $\mathcal{D}^{-1}$ in the expression be implemented the right way around.

So long as the time-sequences in the model are functions from all of the integers, positive and negative, so that there is no first time and no last time, you can show that $\mathcal{D} \, ; \mathcal{D}^{-1} = \iota = \mathcal{D}^{-1} \, ; \mathcal{D}$. There is a perfect symmetry between $\mathcal{D}$ and $\mathcal{D}^{-1}$ as presented here: you should not think of either of them as being more realistic, or more implementable than the other. Indeed, one of them is just the other one seen the other way around.

Even though $\mathcal{D}$ makes no sense in the combinational model, reasoning about circuits that contain $\mathcal{D}$ can be done in exactly the same way as about combinational circuits. Any calculation with a variable R in it, a calculation that does not depend on the particular properties of R, can be instantiated by putting $\mathcal{D}$ for R. For example Horner's rule that if $[R, R] \, ; S = S \, ; R$ then $((\text{tri } R) \setminus apr^{-1}) \, ; \text{rdr } S = \text{rdr}(\text{tri } R \, ; S)$ is a property of the combining forms like composition and reduce right, and not of R or S. Consequently it is the case that if $[\mathcal{D}, \mathcal{D}] \, ; S = S \, ; \mathcal{D}$ then $((\text{tri } \mathcal{D}) \setminus apr^{-1}) \, ; \text{rdr } S = \text{rdr}(\text{tri } \mathcal{D} \, ; S) = \text{rdr}(\text{snd } \mathcal{D} \, ; S)$.

Of course there are things that are true about $\mathcal{D}$ which do depend on its being a particular circuit, for example that $\text{tri } \mathcal{D} \, ; \text{tri } \mathcal{D}^{-1} = \text{map } \iota$, but you should always keep in mind that it is also a perfectly normal component of the kind that we have been reasoning about all along. It is just this simplicity that makes the approach presented here so useful.

One of the remarkable properties of $\mathcal{D}$ is quite how polymorphic it is: $\mathcal{D}$ can be used to represent a unit delay applied to signals of any type, and the tightest type-specification that can be given is that $\mathcal{D} : \iota \to \iota$. Moreover $\mathcal{D} \backslash n = (\text{map }\mathcal{D}) \backslash n$, so that for example $apl\,;\mathcal{D} = apl\,;\text{map }\mathcal{D} = [\mathcal{D}, \text{map }\mathcal{D}]\,;apl$. This means that even though we may be talking about circuit components that will be realized as latches we do not *need* to be concerned with exactly what type of signals they handle, just with the delaying properties. The effect is a clean separation of two concerns: that of making sure that the correct values are handled, and that of ensuring that they arrive at the right times.

1.6.4   Timelessness

For any stateless circuit R you can show that $R = R \backslash \mathcal{D}$. In these notes any circuit R for which $R = R \backslash \mathcal{D}$ is said to be *timeless*. A timeless circuit is one that implements the same relation now as it did in the past, and similarly that it will in the future. Notice that timelessness is *not* the same as statelessness: $\mathcal{D}$ is most definitely not stateless, indeed it the essence of statefulness! Nevertheless since $\mathcal{D} = \mathcal{D}^{-1}\,;\mathcal{D}\,;\mathcal{D} = \mathcal{D} \backslash \mathcal{D}$ it is immediate from the definition that $\mathcal{D}$ is timeless.

Just as a stateless circuit is one that does not behave differently at any time according to the history of its inputs, so a timeless circuit is one that does not behave differently on account of the absolute time. A stateless circuit is necessarily timeless. An example of a timeful circuit would be one that generated a signal indicating time zero. A different sort of example would be E for which $x\,E\,y$ if and only if $\forall i.\,x_{2i} = y_{2i}$. This circuit is not anchored at a single instant because $E \backslash \mathcal{D}^2 = E$, but nevertheless E is not timeless since $E \backslash \mathcal{D} \neq E$.

One of the reasons that timelessness is interesting is that exactly like statelessness it is preserved by all of the circuit constructors that we have seen so far.

1.6.5   Slowing

So far we have ways of talking only about circuits in which all of the sequential components operate at the same rate. Sometimes it proves necessary to reason about circuits in which some components operate at different rates from others. In particular there are various common techniques which require some parts of a circuit to be clocked at twice the rate of other parts, or some such small multiple.

In order to talk about such circuits, we use a new primitive relation *pair* defined by $x\,pair\,y \iff \forall i.\,y_i = \langle x_{2i}, x_{2i+1} \rangle$. Define a 2-slow version of R by $\text{slow }R = [R, R] \backslash pair^{-1}$. The effect is that slow R separates the time-sequences in its domain and range each into two interleaved sequences, and

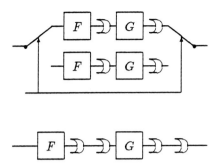

Figure 1.28: slow$(F \;;\; \mathcal{D} \;;\; G \;;\; \mathcal{D})$ and $F \;;\; \mathcal{D}^2 \;;\; G \;;\; \mathcal{D}^2$

performs the calculation described by R on each of the pairs of corresponding subsequences.

It is a matter of proof that if R is an expression involving only the combining forms introduced so far and $\mathcal{D}$ and stateless components, then slow R is the same as the circuit obtained by replacing all occurrences of $\mathcal{D}$ in R by $\mathcal{D}^2$. For example slow $\mathcal{D} = \mathcal{D}^2$, and slow$(F \;;\; \mathcal{D} \;;\; G \;;\; \mathcal{D}) = F \;;\; \mathcal{D}^2 \;;\; G \;;\; \mathcal{D}^2$. The proof is by structural induction on R.

This says that you can implement slow R either by following the recipe in the definition – making two copies of R and supplying them with alternate inputs to derive alternate outputs – or by doubling up each latch in R – effectively sharing the rest of the hardware in R between two calculations, with the state of each calculation being held in the extra latch while the other calculation is being done.

1.6.6 Retiming

Retiming is a procedure for disposing latches through a circuit, with a view to improving properties like the length of the longest unbroken propagation path. The name and the formalisation are usually attributed to Leiserson [Leis81] although the technique is well-known and presentations of it are to be found throughout the literature.

The simplest form of retiming is the replacement of (timeless) R by $R \setminus \mathcal{D}$ or by $R \setminus \mathcal{D}^{-1}$. We can get more complicated retiming laws by applying this transformation independently to different but possibly overlapping parts of a circuit. Horner's rule yields an example of such a composite retiming law. It says that if $S = [\mathcal{D}^{-1}, \mathcal{D}^{-1}] \;;\; S \;;\; \mathcal{D} = (2 \;;\; S) \setminus \mathcal{D}$, that is if S is timeless, then $((\mathsf{tri}\, \mathcal{D}) \setminus apr^{-1}) \;;\; \mathsf{rdr}\, S = \mathsf{rdr}(\mathsf{snd}\, \mathcal{D} \;;\; S)$. A circuit like $\mathsf{rdr}(\mathsf{snd}\, \mathcal{D} \;;\; S)$ is said to be pipelined, because there is a latch between each of the stages of the reduction. You can read this instance of Horner's rule as follows: to implement a $\mathsf{rdr}\, S$ that is pipelined, implement $\mathsf{rdr}(\mathsf{snd}\, \mathcal{D} \;;\; S)$, and adjust the timing of the domain signals according to $(\mathrm{tri}\,\mathcal{D}) \setminus apr^{-1}$. If the circuit takes input from the domain, that adjustment consists of supplying each component of the input one time step earlier than the one to its left.

Such a specification is a *data skew*, and an expression like $(\mathrm{tri}\,\mathcal{D}) \setminus apr^{-1}$ can either represent a skewing circuit which is implemented and performs the required adjustment, or else represent a specification for the interface to the circuit which is implemented.

1.7   A systolic correlator

This section treats a design as an example of the style which we are advocating. Correlation is one of the most important functions in digital signal processing.

1.7.1   Specifying the correlator

The requirement is to calculate

$$c(t) = \sum_{i=0}^{N-1} d(t-i) \times r_i(t)$$

at each time t given time sequences d and r_i.

The first thing to do is to recast the specification into a more tractable notation. We separate the time and space dimensions by introducing a new sequence of values, d'_i, for which

$$c(t) = \sum_{i=0}^{N-1} d'_i(t) \times r_i(t) \quad \& \quad d'_i(t) = d(t-i)$$

the point being that the part with the arithmetic in it looks as though it can be implemented by a stateless circuit calculating

$$c = \sum_{i=0}^{N-1} d'_i \times r_i$$

and the other part looks like a shift-register.

Translating the stateless part of the specification, suppose we have components *mul* and *acc* for which $\langle x, y \rangle \; mul \; (x \times y)$ and $x \; acc \; (\sum_i x_i)$, then

$$\langle d', r \rangle \; (zip \; ; \mathsf{map} \; mul \; ; acc) \; c$$

There is a design decision made in dividing the inputs and outputs between domain and range. It is regrettably hard to modify this particular decision without starting again.

We can also divide the shift register into a stateless and a sequential part by introducing another sequence of values, d_i'', for which

$$d(t) = d_i''(t) \quad \& \quad d_i''(t - i) = d_i'(t)$$

for each i. The first part is stateless, and is a many-way *fork*. Suppose we call this fork F, so that $d \, F \, d''$. From the definition of $\mathcal{D}$, we can rewrite the second part as $d_i'' \, \mathcal{D}^i \, d_i'$, or as $d'' \operatorname{tri} \mathcal{D} \, d'$, so $shift = F \, ; \operatorname{tri} \mathcal{D}$. The whole correlator is implemented as a composition of these parts.

$$\langle d, r \rangle \, (\text{fst } shift \, ; zip \, ; \operatorname{map} mul \, ; acc) \, c$$

Notice that we can implement all of the components as functions from the domain to the range, so in a sense this is a complete design of a correlator. The rest of the development is going to be a matter of optimizing this initial implementation. We need to choose suitable lower level implementations of the many-way fork and the accumulator, so that we can eliminate $\operatorname{tri} \mathcal{D}$ and *zip*, both of which are inefficient in layout area. We also need to translate the implementation from one that operates on numbers to one that operates on binary representations of numbers.

1.7.2 Implementing the shift register

There are many different ways of building a many-way fork from a two-way fork. For example,

$$
\begin{aligned}
F &= \pi_1^{-1} \, ; \operatorname{row}(\pi_1 \, ; fork) \, ; \pi_1 \\
&= \pi_2^{-1} \, ; \operatorname{col}(\pi_2 \, ; fork) \, ; \pi_2
\end{aligned}
$$

Here, we have allowed for the possibility that we may want the list in the range of F to be of zero length. This in turn permits r to be a list of length zero. We will choose to use the version built using row because we know that this circuit is composed with $\operatorname{tri} \mathcal{D}$ and we have a version of Horner's rule that matches.

$$
\begin{aligned}
shift &= \pi_1^{-1} \, ; \operatorname{row}(\pi_1 \, ; fork) \, ; \pi_1 \, ; \operatorname{tri} \mathcal{D} \\
&= \pi_1^{-1} \, ; \operatorname{row}(\pi_1 \, ; fork) \, ; ((\operatorname{tri} \mathcal{D}) \setminus apr^{-1}) \, ; \pi_1 \\
&= \pi_1^{-1} \, ; \operatorname{row}(\pi_1 \, ; fork \, ; \operatorname{snd} \mathcal{D}) \, ; \pi_1
\end{aligned}
$$

The decision to implement the shift register using a row will influence design decisions such as the choice of how to build the accumulator.

1.7.3   Eliminating the *zip*

The next step is to use our results about zipping rows together to eliminate the *zip* from

$$\text{fst } shift \; ; \; zip$$

The components next to the zip are first cast into the form of two rows and these are then interleaved.

$$\text{fst}(\pi_1^{-1} \; ; \; \text{row}(\pi_1 \; ; \; fork \; ; \; \text{snd } \mathcal{D}) \; ; \; \pi_1) \; ; \; zip$$

$\quad = \{\text{type of } zip\}$

$\qquad [\pi_1^{-1} \; ; \; \text{row}(\pi_1 \; ; \; fork \; ; \; \text{snd } \mathcal{D}) \; ; \; \pi_1, \text{map } \iota] \; ; \; zip$

$\quad = \{\text{implementing map } \iota \text{ as a row}\}$

$\qquad [\pi_1^{-1} \; ; \; \text{row}(\pi_1 \; ; \; fork \; ; \; \text{snd } \mathcal{D}) \; ; \; \pi_1, \pi_2^{-1} \; ; \; \text{row}(\pi_2 \; ; \; \pi_1^{-1}) \; ; \; \pi_1] \; ; \; zip$

$\quad = \{\text{rearranging terms}\}$

$\qquad [\pi_1^{-1}, \pi_2^{-1}] \; ; \; [\text{row}(\pi_1 \; ; \; fork \; ; \; \text{snd } \mathcal{D}), \text{row}(\pi_2 \; ; \; \pi_1^{-1})] \; ; \; [\pi_1, \pi_1]' \; ; \; zip$

$\quad = \{ \; [\pi_1^{-1}, \pi_2^{-1}] = [\pi_1^{-1}, \pi_2^{-1}] \; ; \; zip \; \}$

$\qquad [\pi_1^{-1}, \pi_2^{-1}] \; ; \; zip \; ;$

$\qquad\quad [\text{row}(\pi_1 \; ; \; fork \; ; \; \text{snd } \mathcal{D}), \text{row}(\pi_2 \; ; \; \pi_1^{-1})] \; ;$

$\qquad\quad [\pi_1, \pi_1] \; ; \; zip$

$\quad = \{[\pi_1, \pi_1] \; ; \; zip = zip^{-1} \; ; \; \text{fst } zip \; ; \; \pi_1 \}$

$\qquad [\pi_1^{-1}, \pi_2^{-1}] \; ; \; zip \; ;$

$\qquad\quad [\text{row}(\pi_1 \; ; \; fork \; ; \; \text{snd } \mathcal{D}), \text{row}(\pi_2 \; ; \; \pi_1^{-1})] \; ;$

$\qquad\qquad zip^{-1} \; ; \; \text{fst } zip \; ; \; \pi_1$

$\quad = \{\text{interleaving}\}$

$\qquad [\pi_1^{-1}, \pi_2^{-1}] \; ; \; \text{snd } zip \; ; \; \text{row } R \; ; \; \pi_1$

$\qquad\quad \text{where } R = [\pi_1 \; ; \; fork \; ; \; \text{snd } \mathcal{D}, \pi_2 \; ; \; \pi_1^{-1}] \setminus zip^{-1}$

$\quad = \{ \; \pi_2^{-1} \; ; \; zip = \text{map } \pi_2^{-1} \}$

$\qquad [\pi_1^{-1}, \text{map } \pi_2^{-1}] \; ; \; \text{row } R \; ; \; \pi_1$

Now for the interleaved components of which we have a row: the description we have for this cell is one that consists of two separate components and enough plumbing to make it appear that they are interleaved. A more efficient implementation can be obtained by a bit of simplification, using the knowledge that the left-most input to the row is provided through fst π_1^{-1} and the top input through snd map π_2^{-1}.

$$[\pi_1^{-1}, \pi_2^{-1}] \; ; \; R \quad = \quad [\pi_1^{-1}, \pi_2^{-1}] \; ; \; [\pi_1 \; ; \; fork \; ; \; \text{snd } \mathcal{D}, \pi_2 \; ; \; \pi_1^{-1}] \setminus zip^{-1}$$

$$= \quad [\pi_1^{-1}, \pi_2^{-1}] \; ; \; [\pi_1 \; ; \; fork \; ; \; \text{snd } \mathcal{D}, \pi_2 \; ; \; \pi_1^{-1}] \; ; \; zip^{-1}$$

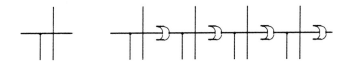

Figure 1.29: $dist = fork$; snd π_1 and row($dist$; snd $\mathcal{D}$)

$$
\begin{aligned}
&= [fork \text{ ; snd } \mathcal{D}, \pi_1{}^{-1}] \text{ ; } zip^{-1} \\
&= fork \text{ ; snd}(\pi_1 \text{ ; } \mathcal{D} \text{ ; } \pi_2{}^{-1}) \\
&= dist \text{ ; snd } \mathcal{D} \text{ ; snd } \pi_2{}^{-1} \\
&\quad \text{where } dist = fork \text{ ; snd } \pi_1
\end{aligned}
$$

so substituting for R in a row R in its context

$$
\begin{aligned}
[\pi_1{}^{-1}, \text{map } \pi_2{}^{-1}] \text{ ; row } R \text{ ; } \pi_1 &= \text{row}(dist \text{ ; snd } \mathcal{D}) \text{ ; snd } \pi_2{}^{-1} \text{ ; } \pi_1 \\
&= \text{row}(dist \text{ ; snd } \mathcal{D}) \text{ ; } \pi_1
\end{aligned}
$$

and putting together what we have so far

$$
CORR = \text{row}(dist \text{ ; snd } \mathcal{D}) \text{ ; } \pi_1 \text{ ; map } mul \text{ ; } acc
$$

implements a relation for which $\langle r, d \rangle\ CORR\ c$.

Since we know that acc can be implemented by rows or columns of components which add a pair of numbers, it looks as though we might be able to implement $CORR$ as a pair of rows, one below the other. The point of doing this is that $(\text{row } S) \updownarrow (\text{row } T) = \text{row}(S \updownarrow T)$ and we can hope to make further simplifications in the component $S \updownarrow T$.

1.7.4 Implementing the accumulator

On the face of it, there is no choice here since we want to implement acc as a row, it will have to be something like a left reduction of add, say $acc = apl^{-1}$; rdl add. This would be a reasonable implementation, but since it does not allow empty rows it would not be natural in the context of a potentially zero-width shift register. Define instead

$$
\begin{aligned}
acc &= \pi_2{}^{-1} \text{ ; fst } zero \text{ ; rdl } add \\
&\quad \text{where } x\ zero\ y \iff x = y = 0 \\
&= \pi_2{}^{-1} \text{ ; fst } zero \text{ ; row}(add \text{ ; } \pi_2{}^{-1}) \text{ ; } \pi_2
\end{aligned}
$$

and if we do indeed proceed in this direction

$$
\begin{aligned}
CORR &= \text{row}(dist \text{ ; snd } \mathcal{D}) \text{ ; } \pi_1 \text{ ; map } mul \text{ ; } \\
&\quad \pi_2{}^{-1} \text{ ; fst } zero \text{ ; row}(add \text{ ; } \pi_2{}^{-1}) \text{ ; } \pi_2
\end{aligned}
$$

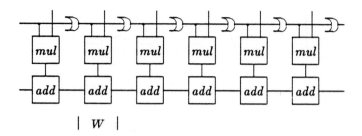

Figure 1.30: an instance of rdl(W ; π_2 ; snd $\mathcal{D}$)

$$
\begin{aligned}
= \quad & \text{row}(\textit{dist} \; ; [\textit{mul}, \mathcal{D}]) \; ; \pi_1 \; ; \\
& \pi_2{}^{-1} \; ; \text{fst } \textit{zero} \; ; \text{row}(\textit{add} \; ; \pi_2{}^{-1}) \; ; \pi_2 \\
= \quad & \text{fst } \pi_2{}^{-1} \; ; \\
& ((\text{fst } \textit{zero} \; ; \text{row}(\textit{add} \; ; \pi_2{}^{-1})) \updownarrow \text{row}(\textit{dist} \; ; [\textit{mul}, \mathcal{D}])) \; ; \\
& \text{snd } \pi_1 \; ; \pi_2 \\
= \quad & \text{fst}(\pi_2{}^{-1} \; ; \text{fst } \textit{zero}) \; ; \\
& \text{row}((\textit{add} \; ; \pi_2{}^{-1}) \updownarrow (\textit{dist} \; ; [\textit{mul}, \mathcal{D}])) \; ; \\
& \pi_2 \; ; \pi_1 \\
= \quad & \text{fst}(\pi_2{}^{-1} \; ; \text{fst } \textit{zero}) \; ; \text{row}(W \; ; \text{snd snd } \mathcal{D}) \; ; \pi_2 \; ; \pi_1 \\
& \text{where } W = (\textit{add} \; ; \pi_2{}^{-1}) \updownarrow (\textit{dist} \; ; \text{fst } \textit{mul}) \\
= \quad & \text{fst}(\pi_2{}^{-1} \; ; \text{fst } \textit{zero}) \; ; \text{rdl}(W \; ; \pi_2 \; ; \text{snd } \mathcal{D}) \; ; \pi_1
\end{aligned}
$$

This completes the rearrangement of the correlator into a very regular circuit
with few different sorts of components, and only local connections between
those components. In fact there is only one sort of component other than
latches: the cell W performs the only 'word-level' operation that we need.
By 'word' here we just mean a number: we are making a distinction with
'bit-level' operations which deal only with ones and noughts.

1.7.5   Making the circuit systolic

Unfortunately although the wires explicit in the structure of the row are
all short – which is what we mean by saying that the connections are all
local – there are some potentially long propagation paths in the circuit. The
one which stands out in figure 1.30 is the lower path along the row, that
through the accumulator. We would prefer to avoid these long paths, on the
assumption that they will take a long time to settle into a known state. Our
strategy is to trade latency for throughput by pipelining the long paths.

Yet another instance of Horner's rule can be used to pipeline this path.

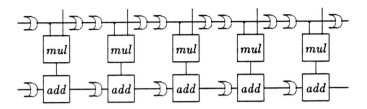

Figure 1.31: an instance of $\mathsf{rdl}(\mathsf{fst}\,\mathcal{D}\;;\,W\;;\,\pi_2\;;\,\mathsf{snd}\,\mathcal{D})$

Knowing that $((\mathsf{irt}\,\mathcal{D})\setminus apl^{-1})\;;\,\mathsf{rdl}\,R = \mathsf{rdl}(\mathsf{fst}\,\mathcal{D}\;;\,R)$ tempts us to adjust the timing of the inputs and implement not $CORR$ but

$$((\mathsf{irt}\,\mathcal{D})\setminus apl^{-1})\;;\,CORR$$
$$= \;(\mathsf{irt}\,\mathcal{D})\setminus apl^{-1}\;;\,\mathsf{fst}(\pi_2^{-1}\;;\,\mathsf{fst}\,zero)\;;\,\mathsf{rdl}(W\;;\,\pi_2\;;\,\mathsf{snd}\,\mathcal{D})\;;\,\pi_1$$
$$= \;\mathsf{fst}(\pi_2^{-1}\;;\,\mathsf{fst}\,zero)\;;\,((\mathsf{irt}\,\mathcal{D})\setminus apl^{-1})\;;\,\mathsf{rdl}(W\;;\,\pi_2\;;\,\mathsf{snd}\,\mathcal{D})\;;\,\pi_1$$
$$= \;\mathsf{fst}(\pi_2^{-1}\;;\,\mathsf{fst}\,zero)\;;\,\mathsf{rdl}(\mathsf{fst}\,\mathcal{D}\;;\,W\;;\,\pi_2\;;\,\mathsf{snd}\,\mathcal{D})\;;\,\pi_1$$

an example of which is shown in figure 1.31. This circuit has no combinational propagation paths longer than one cell. Such a design is called systolic, in this case word-level systolic, perhaps because the clocking of the latches pumps data around the circuit. As we use it here the word 'systolic' means just that there are no combinational paths (at a given level of detail).

The problem with this implementation of the correlator is that, contrary to appearances in our diagrams, latches tend to be much bigger and so more expensive than the other components in a circuit. The design in figure 1.31 has in each cell three latches two of which are big enough to store a value of d, and one of which is big enough to store a value of the accumulated sum.

Looking for an alternative strategy we might try skewing in the other direction

$$((\mathsf{irt}\,\mathcal{D}^{-1})\setminus apl^{-1})\;;\,CORR$$
$$= \;\mathsf{fst}(\pi_2^{-1}\;;\,\mathsf{fst}\,zero)\;;\,\mathsf{rdl}(\mathsf{fst}\,\mathcal{D}^{-1}\;;\,W\;;\,\pi_2\;;\,\mathsf{snd}\,\mathcal{D})\;;\,\pi_1$$

which introduces anti-delays into the accumulator. In order to be able to implement these by latches we would need to reverse the data-flow in the accumulator. This would not be too difficult to do. However, the real problem with this implementation is that 'unskewing' has cancelled the effect of the latches in the shift register, for

$$\mathsf{rdl}(\mathsf{fst}\,\mathcal{D}^{-1}\;;\,W\;;\,\pi_2\;;\,\mathsf{snd}\,\mathcal{D})$$
$$= \;\mathsf{fst}\,\mathcal{D}^{-1}\;;\,\mathsf{rdl}(W\;;\,\pi_2\;;\,\mathsf{snd}\,\mathcal{D}\;;\,\mathcal{D}^{-1})\;;\,\mathcal{D}$$
$$= \;\mathsf{fst}\,\mathcal{D}^{-1}\;;\,\mathsf{rdl}(W\;;\,\pi_2\;;\,\mathsf{fst}\,\mathcal{D}^{-1}\;;\,\mathsf{snd}(\mathcal{D}\;;\,\mathcal{D}^{-1}))\;;\,\mathcal{D}$$
$$= \;\mathsf{fst}\,\mathcal{D}^{-1}\;;\,\mathsf{rdl}(W\;;\,\pi_2\;;\,\mathsf{fst}\,\mathcal{D}^{-1})\;;\,\mathcal{D}$$

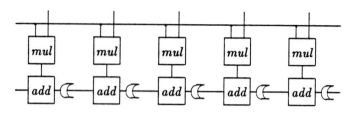

Figure 1.32: an instance of rdl(W ; π_2 ; fst $\mathcal{D}^{-1}$)

and there is now a long combinational propagation path along the top of the circuit, as illustrated in figure 1.32.

We have run up against a common problem in regular array design. We have chosen directions of data flow in such a way that attempting to make the circuit systolic results either in too many latches or in long combinational paths. The solution is to reverse the direction of one of the data paths and then to slow the circuit before skewing. Doubling up the delays in the shift register ensures that skewing only wipes out half of those delays, so that both paths end up being properly pipelined.

We choose to reverse the direction of the shift-register because that looks easier. Since swp row R = snd rev ; row swp R ; fst rev and swp($dist$; snd $\mathcal{D}$) = fst $\mathcal{D}^{-1}$; $dist$ we can calculate that

$$\langle d, r \rangle \text{ (row}(dist \text{ ; snd } \mathcal{D}) \text{ ; } \pi_1) \text{ } s$$
$$\Longleftrightarrow \quad r \left(\pi_2{}^{-1} \text{ ; swp row}(dist \text{ ; snd } \mathcal{D}) \right) \langle s, d \rangle$$
$$\Longleftrightarrow \quad r \left(rev \text{ ; } \pi_2{}^{-1} \text{ ; row}(\text{fst } \mathcal{D}^{-1} \text{ ; } dist) \text{ ; fst } rev \right) \langle s, d \rangle$$

We know from our first design that s (map mul ; acc) c, so

$$r \left(rev \text{ ; } \pi_2{}^{-1} \text{ ; row}(\text{fst } \mathcal{D}^{-1} \text{ ; } dist) \text{ ; fst}(rev \text{ ; map } mul \text{ ; } acc) \right) \langle c, d \rangle$$

and because rev ; map mul = map mul ; rev and rev ; acc = acc we can remove the internal rev. Clearly we do not want to implement the rev on the left-hand end in silicon, so we play the trick of changing the way in which the inputs are presented to the circuit, and implementing rev ; $CORR'$ where r $CORR'$ $\langle c, d \rangle$.

$$rev \text{ ; } CORR' \quad = \quad \pi_2{}^{-1} \text{ ; row}(\text{fst } \mathcal{D}^{-1} \text{ ; } dist) \text{ ; fst}(\text{map } mul \text{ ; } acc)$$
$$= \quad \pi_2{}^{-1} \text{ ; row}(\text{fst } \mathcal{D}^{-1} \text{ ; } dist \text{ ; fst } mul) \text{ ; }$$
$$\text{fst}(\pi_2{}^{-1} \text{ ; fst } zero \text{ ; row}(add \text{ ; } \pi_2{}^{-1}) \text{ ; } \pi_2)$$

As before, we rearrange this into a pair of rows, one below the other.

$$rev \text{ ; } CORR' \quad = \quad \pi_2{}^{-1} \text{ ; fst fst } zero \text{ ; row}(\text{fst snd } \mathcal{D}^{-1} \text{ ; } W) \text{ ; } \pi_2$$
$$\text{where } W = (add \text{ ; } \pi_2{}^{-1}) \updownarrow (dist \text{ ; fst } mul)$$
$$= \quad \pi_2{}^{-1} \text{ ; fst fst } zero \text{ ; rdl}(\text{fst snd } \mathcal{D}^{-1} \text{ ; } W \text{ ; } \pi_2)$$

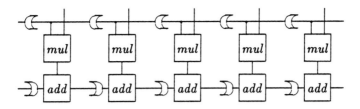

Figure 1.33: an instance of $\mathsf{rdl}(\mathsf{fst}[\mathcal{D}, \mathcal{D}^{-1}] \; ; W \; ; \pi_2)$

Next, we slow the circuit

$$
\begin{aligned}
rev \; ; \mathsf{slow}\, CORR' &= \mathsf{slow}(rev \; ; CORR') \\
&= {\pi_2}^{-1} \; ; \mathsf{fst}\, \mathsf{fst}\, zero \; ; \mathsf{rdl}(\mathsf{fst}\, \mathsf{snd}\, \mathcal{D}^{-2} \; ; W \; ; \pi_2)
\end{aligned}
$$

and retime using Horner's rule

$$
\begin{aligned}
\mathsf{irt}\, \mathcal{D} \; ; rev \; &; \mathsf{slow}\, CORR' \\
&= \mathsf{irt}\, \mathcal{D} \; ; {\pi_2}^{-1} \; ; \mathsf{fst}\, \mathsf{fst}\, zero \; ; \mathsf{rdl}(\mathsf{fst}\, \mathsf{snd}\, \mathcal{D}^{-2} \; ; W \; ; \pi_2) \\
&= {\pi_2}^{-1} \; ; \mathsf{fst}\, \mathsf{fst}\, zero \; ; ((\mathsf{irt}\, \mathcal{D}) \setminus apl^{-1}) \; ; \mathsf{rdl}(\mathsf{fst}\, \mathsf{snd}\, \mathcal{D}^{-2} \; ; W \; ; \pi_2) \\
&= {\pi_2}^{-1} \; ; \mathsf{fst}\, \mathsf{fst}\, zero \; ; \mathsf{rdl}(\mathsf{fst}(\mathcal{D} \; ; \mathsf{snd}\, \mathcal{D}^{-2}) \; ; W \; ; \pi_2) \\
&= {\pi_2}^{-1} \; ; \mathsf{fst}\, \mathsf{fst}\, zero \; ; \mathsf{rdl}(\mathsf{fst}[\mathcal{D}, \mathcal{D}^{-1}] \; ; W \; ; \pi_2)
\end{aligned}
$$

which is illustrated in figure 1.33. The left-hand side reminds us that we have only implemented a slow correlator and that the weights r should be presented to the circuit in reverse order and time-skewed.

That completes the development of two word-level systolic correlators, one of which uses $3N$ latches, the other of which uses only $2N$ at the expense of a doubled clock speed.

1.7.6 Refining to a bit level implementation

The next stage of the development is to refine the operations to ones that combine and produce bit-vectors, and there we will stop. Earlier presentations of this circuit such as reference [Sheer88] have made this refinement step in one leap, appealing to 'well-known' implementations of the arithmetic functions. However since we can express representation and abstraction relations in the same notation as the circuit, we have the opportunity to bridge the gap between word- and bit-level. This is work still in progress at Glasgow and Oxford, so this section does no more than outline the shape that the refinement would have.

The strategy for refining from word- to bit-level is: to describe an abstraction relation that relates the bit-vectors operated on by the circuit to the numbers that they represent; to compose this relation with the word-level description of the correlator; and then to simplify the resulting expression. We will use a simple bit-parallel binary representation of numbers, most significant bit first. We know to choose this representation because we have done the development before using the least significant bit first representation and regretted the choice; the resulting circuit was hard to pipeline. Choosing the representation with the most-significant-bit first gives simpler data flow through the bit-level implementation of the cell W.

If bit is the identity relation on bits, that is if it is the type of bits, and $bin = \text{map } bit \; ; rev \; ; \text{tri}(\times 2) \; ; acc$, then since $bin^{-1} \; ; bin = nat$ is the identity on natural numbers, bin is an abstraction from bit vectors to the natural numbers. Moreover if $bin_n = n \; ; bin$ then $bin_n \; ; bin_n^{-1} = n \; ; \text{map } bit$ and $bin_n^{-1} \; ; bin_n = nat_n$ where nat_n is the type of those numbers in the range $0 \le i < 2^n$ representable in n bits, a sub-type of nat. So bin_n abstracts n-bit bit-vectors to small enough numbers, and bin_n^{-1} represents any small enough number as an n-bit bit-vector.

Suppose we are not going to want to give negative inputs to the correlator: then we can show that no part of the circuit need then deal with negative numbers, and that the output must be non-negative. The next thing to do is to decide on some appropriate widths for the bit-vectors. Since each of the component parts of W is going to become a column of cells, and since we will want to interleave these columns, it will be simplest if we can arrange for everything to have the same width. This is a great simplification, but one that you need not make if you are determined to produce the smallest possible implementation of the circuit.

As in the circuit in reference [McCab82], we further simplify matters by assuming that the reference inputs r_i are only a single bit wide. If the largest d-value can be represented by a k-wide bit vector – that is, if it is in nat_k – then any value manipulated by the circuit can be represented by an n-wide bit-vector, provided $n \ge N + k$.

$$N \; ; \text{map } nat_1 \; ; \text{irt } \mathcal{D} \; ; rev \; ; \text{slow } CORR' \; ; \text{snd } nat_k$$

$$= \quad N \; ; \text{map } nat_1 \; ; \text{irt } \mathcal{D} \; ; rev \; ; \text{slow } CORR' \; ; [nat_n, nat_k]$$

$$= \quad N \; ; \text{map } nat_1 \; ; \text{irt } \mathcal{D} \; ; rev \; ; \text{slow } CORR' \; ; [nat_n, nat_n] \; ; \text{snd } nat_k$$

Now our strategy is to replace each nat_n by $bin_n^{-1} \; ; bin_n$ and to push the representation relations down into the circuit, until the circuit can be described as a collection of components that we know how to implement. Doing this, we find that we need to implement

$$[[bin_n, bin_n], bit] \; ; W \; ; \pi_2 \; ; [bin_n, bin_n]^{-1}$$

$$= \quad (([bin_n, bin_n] \; ; add \; ; bin_n^{-1} \; ; \pi_2^{-1}) \updownarrow$$

$$(dist \; ; \text{fst}([bin_n, bit] \; ; mul \; ; bin_n^{-1}))) \; ; \pi_2$$

We have shown elsewhere [Jon90] how expressions like $[bin_n, bin_n]$; add; bin_n^{-1} can be implemented by arrays of simpler cells. Here, we state the necessary theorems without proof. Firstly

$$[bin_n, bin_n] ; add ; bin_n^{-1} = zip ; n ; \pi_1^{-1} ; \text{snd } zero ; \text{col } FA ; \pi_2$$

where the relation FA is just a standard full-adder, a bit-level component that we know how to implement:

$$FA = [[bit, bit] ; add, bit] ; add ; add^{-1} ; [(\times 2)^{-1} ; bit^{-1}, bit^{-1}]$$

Similarly

$$[bin_n, bit] ; mul ; bin_n^{-1} = \text{col } M ; \pi_2 ; n$$
$$\text{where } M = fork ; [\pi_2, [bit, bit] ; mul ; bit^{-1}]$$

and the bit-multiplier $[bit, bit]$; mul; bit^{-1} can be implemented by the same electronics as you would use to implement logical 'and'.

That implements the addition and multiplication parts of the cell as two separate columns, as illustrated in figure 1.34. It would be a good idea to eliminate some wire crossings, and to do this it will be necessary to interleave the columns. We start by merging the *dist* at the top of the diagram with the column of multiplier cells.

$$dist ; \text{fst}(\text{col } M ; \pi_2 ; n) = \text{col } M' ; \pi_2 ; n ; zip^{-1}$$
$$\text{where } M' = dist ; \text{fst } M ; lsh$$

Now we are in a position to interleave the column of M' cells with the column of FA cells.

$$[[bin_n, bin_n], bit] ; W ; \pi_2 ; [bin_n, bin_n]^{-1}$$
$$= (zip ; n ; \pi_1^{-1} ; \text{snd } zero ; \text{col } FA ; \pi_2 ; \pi_2^{-1}) \updownarrow$$
$$(\text{col } M' ; \pi_2 ; n ; zip^{-1})) ; \pi_2$$
$$= [zip ; n, \pi_1^{-1} ; \text{snd } zero] ; \text{col } B ; \pi_2 ; (zip ; n)^{-1}$$
$$\text{where } B = ((swap \updownarrow M') ; \text{snd } rsh) \leftrightarrow (FA \updownarrow swap)$$

We have stated without justification a theorem about the interleaving of two columns. To approach this design properly, it would have been better to introduce a new higher order function corresponding to the way the cell B is built from the cells FA and M' and to develop some of the algebra of that higher order function.

Figure 1.35 shows the column that we have just made. The relation B can be implemented by a full-adder, an 'and' gate and some wiring; so we can stop here and plug everything back together.

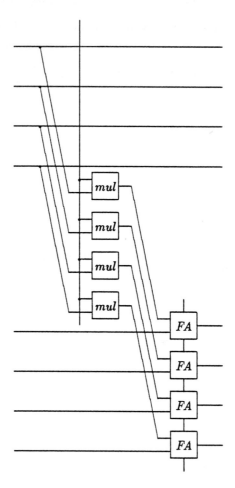

Figure 1.34: $[[bin_4, bin_4], bit]$; W ; $\mathsf{snd}[bin_4, bin_4]^{-1}$ implemented by an interleaving of separate columns

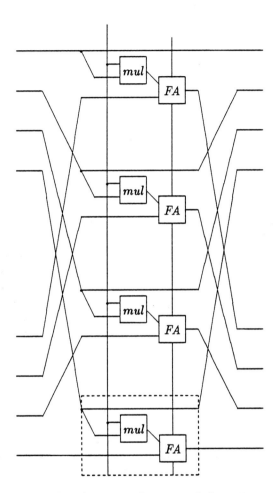

Figure 1.35: $[[bin_4, bin_4], bit] ; W ; snd[bin_4, bin_4]^{-1}$ implemented by a single column of interleaved components

N ; map nat_1 ; irt $\mathcal{D}$; rev ; slow $CORR'$; snd nat_k

$$
\begin{aligned}
= \quad & N \text{ ; map } bit^{-1} \text{ ; } \pi_2{}^{-1} \text{ ; fst fst map } zero \text{ ;} \\
& \text{rdl(fst}[\mathcal{D}, \mathcal{D}^{-1}] \text{ ; } [[bin_n, bin_n], bit] \text{ ; } W \text{ ; } \pi_2 \text{ ; } [bin_n, bin_n]^{-1}) \text{ ;} \\
& [bin_n, bin_n] \text{ ; snd } nat_k \\
= \quad & N \text{ ; map } bit^{-1} \text{ ; } \pi_2{}^{-1} \text{ ; fst fst map } zero \text{ ;} \\
& \text{rdl}([[\mathcal{D}, \mathcal{D}^{-1}] \text{ ; } zip \text{ ; } n, \pi_1{}^{-1} \text{ ; snd } zero] \text{ ; col } B \text{ ; } \pi_2 \text{ ; } (zip \text{ ; } n)^{-1}) \text{ ;} \\
& [bin_n, bin_n \text{ ; } nat_k] \\
= \quad & N \text{ ; map } bit^{-1} \text{ ; } \pi_2{}^{-1} \text{ ; } [zip \text{ ; map fst } zero, \text{map}(\pi_1{}^{-1} \text{ ; snd } zero)] \text{ ;} \\
& \text{rdl(fst map}[\mathcal{D}, \mathcal{D}^{-1}] \text{ ; col } B \text{ ; } \pi_2) \text{ ;} \\
& zip^{-1} \text{ ; } [bin_n, bin_n \text{ ; } nat_k] \\
= \quad & N \text{ ; map } bit^{-1} \text{ ; } \pi_2{}^{-1} \text{ ; } [\text{map fst } zero, \text{map}(\pi_1{}^{-1} \text{ ; snd } zero)] \text{ ;} \\
& \text{row col(fst}[\mathcal{D}, \mathcal{D}^{-1}] \text{ ; } B) \text{ ;} \\
& \pi_2 \text{ ; } zip^{-1} \text{ ; } [bin_n, bin_n \text{ ; } nat_k]
\end{aligned}
$$

The middle term on the right-hand side is just an N by n grid of identical cells, and this part would certainly be implemented as part of the circuit. The terms at either end of the expression are what remains of the representation relation. They describe how the input must be presented to the circuit, and how to read the output: N;map bit^{-1} means 'a list of N one-bit numbers'; and zip^{-1}; $[bin_n, bin_n \text{ ; } nat_k]$ means 'the interleaving of an n-bit binary representation of one number and the n-bit representation of another number which could have been represented in k bits'. The remaining terms describe the tying of certain lines to a level representing the bit zero, and the discarding of the signals on other lines.

1.7.7   Making the implementation systolic at the bit level

The circuit represented by $G = \text{row col(fst}[\mathcal{D}, \mathcal{D}^{-1}] \text{ ; } B)$, illustrated in figure 1.36, has no long horizontal combinational paths because of the $[\mathcal{D}, \mathcal{D}^{-1}]$ through which each horizontal data-path passes at each cell. Moreover since the change of representation that we used did not involve any change in the direction of the data-flow, both the delays and anti-delays in the circuit remain implementable as latches.

There are however vertical unbroken combinational paths through each column of cells. In figure 1.35 these are the long vertical line that carries the reference bit, and a route through the carry-path that joins the FA components. We must eliminate these long paths by skewing. We leave this task as an exercise for the reader. (Hint: the appropriate skewing law for columns is reminiscent of Horner's rule for right reductions:

$$
(\text{tri } \mathcal{D} \setminus apr^{-1}) \text{ ; col } R \text{ ; snd tri } \mathcal{D}^{-1} = \text{col(snd } \mathcal{D} \text{ ; } R)
$$

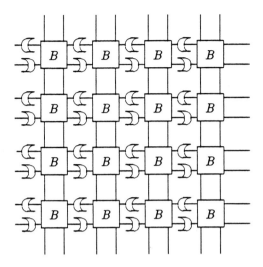

Figure 1.36: an instance of row col(fst$[\mathcal{D}, \mathcal{D}^{-1}]$; B)

and although we could skew each column in turn, it is simplest to exchange the row and column structure.)

1.8 Butterfly networks

Butterfly networks and algorithms are a design cliché in digital signal processing, just as arrays and grids are. Typical applications are sorting, fast Fourier transform, and the interconnection of processors and memories or of networks of processors. Both the 'cube connected cycle' and the 'hypercube' can be made by folding up a butterfly network. From the point of view of an algorithm designer, butterfly networks are interesting because they have many simple recursive decompositions. We need some extra Ruby notation in order to be able to describe butterflies.

1.8.1 The perfect shuffle

We have already made a lot of use of the wiring relation *zip* which was defined by $\langle x, y \rangle$ *zip* $z \iff \forall i. z_i = \langle x_i, y_i \rangle$. Butterfly networks use a related piece of plumbing, the perfect shuffle or *riffle*. Consider a deck of fifty-two cards; it can be riffled by dividing in half, interleaving to give twenty-six pairs, and then 'unpairing' by forgetting about the pairing.

This permutation on even-length lists is actually used on silicon despite the crossovers that it introduces. We can describe it by composing three simpler

wiring relations, *halve*, *zip* and *pair*$^{-1}$ ('unpair'). The relation *halve* relates
a list of even length to a pair containing the first and second halves of the
list so $2n$; *halve* $=$ *app*$^{-1}$; $[n, n]$. The relation *pair* defined by x *pair* y $\Longleftrightarrow$
$\forall i.\ y_i = \langle x_{2i}, x_{2i+1} \rangle$ divides a list of length $2n$ into an n-list of pairs. (We used
it earlier when defining slow.) Define

$$riffle \quad = \quad halve\ ;\ zip\ ;\ pair^{-1}$$

Some of the useful properties of *riffle* are

$$
\begin{aligned}
2n\ ;\ riffle &= riffle\ ;\ 2n \\
2n\ ;\ riffle\ ;\ riffle^{-1} &= 2n \\
2n\ ;\ riffle^{-1}\ ;\ riffle &= 2n \\
2^k\ ;\ riffle^k &= 2^k
\end{aligned}
$$

1.8.2   Two and interleave

Although we could now describe butterflies using the combining forms and
wiring relations already introduced, two new combining forms simplify the
description.

Let R and S both be of type map $\iota \to$ map ι. Define

$$R \mid S \quad = \quad [R, S] \setminus halve^{-1}$$

and

$$two\ R \quad = \quad R \mid R$$

The relation two R relates by R the first halves of the lists in its domain and
range, and similarly the second halves. You can think of it as placing two
copies of R across a bus. Similarly two^n R places 2^n copies of R across a bus.

The relation ilv R on the other hand relates by R the even numbered
elements of the lists in its domain and range, and similarly the odd numbered
elements.

$$ilv\ R \quad = \quad (two\ R) \setminus riffle$$

Surprisingly enough, two and ilv can be exchanged

$$two\ ilv\ R \quad = \quad ilv\ two\ R$$

a fact that we will find very useful later. If we know that R and S relate
only lists of equal length, then both ilv and two distribute over composition:
$ilv(R\ ;\ S) = (ilv\ R)\ ;\ (ilv\ S)$ and $two(R\ ;\ S) = (two\ R)\ ;\ (two\ S)$.

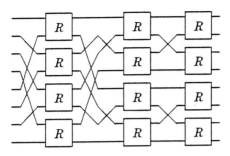

Figure 1.37: a butterfly of size 3

1.8.3   Fat composition

A butterfly network of size k consists of a composition of k ranks, each of which contains the same number of copies of the basic cell, but disposed differently across the bus. We need a way of describing this kind of composition of several different but similar relations. This version of composition, $\mathring{,}$ (read 'fat composition'), when applied to a list of relations composes them. This means that we need a notation for lists: write $\langle i : 0 \leq i < n : x_i \rangle$ for the list of length n, the ith element of which is x_i. So for example $\mathring{,}\langle i : 0 \leq i < 4 : R_i \rangle$ is the relation $R_0 ; R_1 ; R_2 ; R_3$. You can think of $\mathring{,}$ as the quantifier corresponding to composition, just as $\sum$ is the quantifier corresponding to addition.

1.8.4   Describing the butterfly network

Let us assume that our basic cell R is of type $2^{n+1} \to 2^{n+1}$ for some fixed n. Usually n will be zero, for example when we build butterflies of comparators or of two way switches. However making n a parameter will be a useful generalization when trying to understand the recursive structure of the network.

A butterfly network of size k consists of k columns or ranks, each containing 2^{k-1} copies of the basic cell. Each rank distributes these copies across the bus – which must be of width 2^{k+n} – in a different but regular way.

$$2^{n+k} ; \bowtie R \quad = \quad \mathring{,}\langle i : 0 \leq i < k : \text{two}^i \, \text{ilv}^{k-i-1} \, R \rangle$$

A butterfly of size zero is simply the identity on lists of length 2^n. If we fix n to be zero so that $R : 2 \to 2$ then a butterfly of size 3 has 3 ranks, each with 4 copies of R. It can be variously described as

$$2^3 ; \bowtie R \quad = \quad \text{ilv}^2 \, R \, ; \text{two}(\text{ilv} \, R \, ; \text{two} \, R)$$
$$= \quad \text{ilv}^2 \, R \, ; \text{two} \, \text{ilv} \, R \, ; \text{two}^2 \, R$$

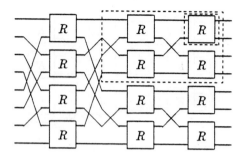

Figure 1.38: first recursive decomposition of the butterfly

$$= \quad \mathsf{ilv}^2\, R \;;\; \mathsf{ilv\,two}\, R \;;\; \mathsf{two}^2\, R$$
$$= \quad \mathsf{ilv(ilv}\, R \;;\; \mathsf{two}\, R) \;;\; \mathsf{two}^2\, R$$

You can think of these descriptions as different ways of decomposing the same relation. Figure 1.37 shows one possible way of laying out the relation. There are of course many others.

Starting from our iterative description of the butterfly, we can use the properties of two, ilv and *riffle* to derive several alternative decompositions. Unwinding the definition of the butterfly from the left gives

$$
\begin{aligned}
2^{n+k+1} \;;\; \bowtie R &= \; {}_9^o\langle i : 0 \le i < k+1 : \mathsf{two}^i\, \mathsf{ilv}^{(k+1)-i-1}\, R\rangle \\
&= \; (\mathsf{ilv}^k\, R) \;;\; {}_9^o\langle i : 1 \le i < k+1 : \mathsf{two}^i\, \mathsf{ilv}^{(k+1)-i-1}\, R\rangle \\
&= \; (\mathsf{ilv}^k\, R) \;;\; {}_9^o\langle j : 0 \le j < k : \mathsf{two}^{j+1}\, \mathsf{ilv}^{(k+1)-(j+1)-1}\, R\rangle \\
&= \; (\mathsf{ilv}^k\, R) \;;\; {}_9^o\langle j : 0 \le j < k : \mathsf{two\,two}^j\, \mathsf{ilv}^{k-j-1}\, R\rangle \\
&= \; (\mathsf{ilv}^k\, R) \;;\; \mathsf{two}\, {}_9^o\langle j : 0 \le j < k : \mathsf{two}^j\, \mathsf{ilv}^{k-j-1}\, R\rangle \\
&= \; (\mathsf{ilv}^k\, R) \;;\; \mathsf{two}(2^{n+k} \;;\; \bowtie R)
\end{aligned}
$$

showing that a butterfly of size $k+1$ is made from 2^k copies of R making up the first rank, some wiring, and two recursive instances of a butterfly of size k. This is perhaps a more familiar description of the butterfly. Figure 1.38 outlines an instance of a butterfly of size 2 and one of size 1 within a butterfly of size 3.

Similarly unwinding the fat composition from the right gives us

$$
\begin{aligned}
2^{n+k+1} \;;\; \bowtie R &= \; {}_9^o\langle i : 0 \le i < k+1 : \mathsf{two}^i\, \mathsf{ilv}^{(k+1)-i-1}\, R\rangle \\
&= \; {}_9^o\langle i : 0 \le i < k : \mathsf{two}^i\, \mathsf{ilv}^{(k+1)-i-1}\, R\rangle \;;\; \mathsf{two}^k\, R \\
&= \; {}_9^o\langle i : 0 \le i < k : \mathsf{ilv\,two}^i\, \mathsf{ilv}^{k-i-1}\, R\rangle \;;\; \mathsf{two}^k\, R \\
&= \; \mathsf{ilv}({}_9^o\langle i : 0 \le i < k : \mathsf{two}^i\, \mathsf{ilv}^{k-i-1}\, R\rangle) \;;\; \mathsf{two}^k\, R \\
&= \; \mathsf{ilv}(2^{n+k} \;;\; \bowtie R) \;;\; \mathsf{two}^k\, R
\end{aligned}
$$

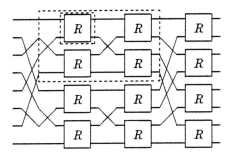

Figure 1.39: a second recursive decomposition

This time we can see that a butterfly consists of two smaller butterflies interleaved, composed with a rank consisting just of two^k R. Figure 1.39 outlines this second recursive decomposition.

Because two and ilv can be exchanged, it must be the case that

$$
\begin{aligned}
\text{two}(2^{n+k} \, ; \bowtie R) &= \text{two} \, \S\langle j : 0 \le j < k : \text{two}^j \, \text{ilv}^{k-j-1} \, R\rangle \\
&= \S\langle j : 0 \le j < k : \text{two two}^j \, \text{ilv}^{k-j-1} \, R\rangle \\
&= \S\langle j : 0 \le j < k : \text{two}^j \, \text{ilv}^{k-j-1}(\text{two } R)\rangle \\
&= 2^{n+k+1} \, ; \bowtie \text{two } R
\end{aligned}
$$

and

$$
\text{ilv}(2^{n+k} \, ; \bowtie R) = 2^{n+k+1} \, ; \bowtie \text{ilv } R
$$

which gives us another pair of decompositions of a butterfly each containing only a single instance of the next smaller butterfly, although with a larger component.

$$
\begin{aligned}
2^{n+k+1} \, ; \bowtie R &= (\text{ilv}^k \, R) \, ; (\bowtie \text{two } R) \\
&= (\bowtie \text{ilv } R) \, ; (\text{two}^k \, R)
\end{aligned}
$$

The second of these decompositions is illustrated in figure 1.40: the larger outlined box is an instance of $\bowtie$ ilv R; and the smaller is just ilv^2 R which is a degenerate instance of $\bowtie$ ilv^2 R.

The butterfly has many more recursive decompositions because we need not divide it into a single rank and a recursive call: it can be the composition of two similar parts. The discovery of this decomposition is left as an exercise for the reader.

Finally, let us return to the case in which the basic cell is of type $2 \to 2$. In that case, we can relate two and ilv to map to get simpler descriptions of the circuit. Suppose that $R : 2 \to 2$, then

$$
2^{k+1} \, ; \text{two}^k \, R = 2^{k+1} \, ; ((\text{map } R) \setminus pair^{-1})
$$

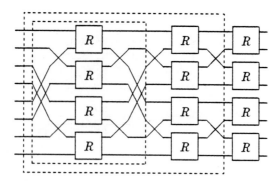

Figure 1.40: singly recursive view

In other words, each R cell operates on pairs of adjacent elements from the lists in the domain and range. The form $((\text{map } R) \setminus pair^{-1})$ arises so often that we abbreviate it to $\text{pmap } R$. Similarly

$$2^{k+1} ; \text{ilv}^k R \;\; = \;\; 2^{k+1} ; ((\text{pmap } R) \setminus riffle^{-1})$$

and we often abbreviate $((\text{pmap } R) \setminus riffle^{-1})$ as $\text{rpmap } R$. The recursive descriptions of a butterfly of $R : 2 \to 2$ can be rewritten as

$$2^{k+1} ; \bowtie R \;\; = \;\; \text{rpmap } R ; \text{two}(2^k ; \bowtie R)$$
$$= \;\; \text{ilv}(2^k ; \bowtie R) ; \text{pmap } R$$

1.9 The Fourier transform

This outlines a development of a common digital signal-processing algorithm, having the aim of turning the specification into the Ruby description of a butterfly circuit. There is a fuller presentation of the calculation of the implementation from the specification in reference [Jon89].

1.9.1 The discrete Fourier transform

Twenty-five years ago Cooley and Tukey rediscovered an optimizing technique usually attributed to Gauss, who used it in hand calculation. They applied the technique to the discrete Fourier transform, reducing an apparently $O(n^2)$ problem to the almost instantly ubiquitous $O(n \log n)$ 'fast Fourier transform' [Cool65].

The fast Fourier transform is not of course a different transform, but a fast implementation of the discrete transform. Its greatest virtue lies in that it can be executed in $O(\log n)$ time on $O(n)$ processors in a uniform way – it lends itself to a low-latency high-throughput pipelined hardware implementation.

The discrete Fourier transform is defined in terms of the arithmetic on an integral domain. You can think of arithmetic on complex numbers, for a definite example, although there are applications where finite fields or vector spaces over integral domains are appropriate. The derivation depends only on the algebraic properties of the arithmetic, not on the underlying arithmetic itself, so everything said here about the algorithm will be true for finite fields and vector spaces as well.

The discrete Fourier transform of a vector x of length n is a vector y of the same length for which

$$y_j \;=\; \sum_{k:0\leq k<n} \omega^{j\times k} \times x_k$$

where ω is a principal n-th root of unity. (In the example of complex numbers, you can think of $\omega = e^{2\pi i/n}$.) The result, y, is sometimes called the 'frequency spectrum' of the sample x.

Even if the powers of ω are pre-calculated, it would appear that $O(n^2)$ multiplications are required to evaluate the whole of y for any x. The fast algorithm avoids many of these by making use of the fact that $\omega^n = 1$. If n is composite, the calculation can be divided into what amounts to a number of smaller Fourier transforms. Suppose $n = p \times q$, then by a change of variables

$$
\begin{aligned}
y_{pa+b} \;&=\; \sum_{c:0\leq c<p}\ \sum_{d:0\leq d<q} \omega^{(pa+b)(qc+d)} x_{qc+d} \\
&=\; \sum_{c:0\leq c<p}\ \sum_{d:0\leq d<q} (\omega^{pq})^{ac}(\omega^{p})^{ad}(\omega^{q})^{bc}\omega^{bd} x_{qc+d} \\
&=\; \sum_{d:0\leq d<q} (\omega^{p})^{ad}\omega^{bd} \sum_{c:0\leq c<p} (\omega^{q})^{bc} x_{qc+d}
\end{aligned}
$$

Since ω^q is a p-th root of unity, and ω^p is a q-th root of unity, it is not surprising that the above calculation leads to an implementation in which p-sized and q-sized transforms appear. It is harder, however, to see what that implementation might be.

1.9.2 Casting the algorithm in the notation

The first task in a calculation dealing with an algorithm is to cast the specification in the notation that will be used to handle the development. There are two things which we do in this stage.

One part appears to be largely a process of eliminating subscripts, since the usual convention is to specify separately each co-ordinate of an output vector.

The conventional understanding of a specification of the form $y_i = \ldots$ is that the subscript is universally quantified, so that this one equation formally represents a number of different equations, one for each value of i. To make clear that an algorithm operates uniformly at all co-ordinates of its output we write a single equation which defines the whole list of output values.

The other part of the translation is to manipulate the specification – which is usually an expression describing the output of a calculation for a given input – into the form of an application to that input of an expression representing the algorithm. The manipulation of the algorithm can then proceed without reference to the particular input.

The discrete Fourier transform was specified by

$$y_j \;=\; \sum_{k:0\le k<n} \omega^{j\times k} \times x_k$$

by which was meant that the output y should be defined for each j in the range $0 \le j < n$, so meaning that

$$y = \langle j : 0 \le j < n : \sum \langle k : 0 \le k < n : \omega^{j\times k} \times x_k \rangle\rangle$$

$\Longleftarrow$ { meaning of summation and map, meaning of exponentiation }

$\qquad \langle j : 0 \le j < n : \langle k : 0 \le k < n : \omega^{j\times k} \times x_k \rangle\rangle\,(\mathsf{map}\,acc)\,y$

$\qquad$ where $acc = apl^{-1}\,;\mathsf{rdl}\,add = apr^{-1}\,;\mathsf{rdr}\,add$

$\Longleftarrow$ { meaning of arithmetic exponentiation, associativity of $\times$ }

$\qquad \langle j : 0 \le j < n : \langle k : 0 \le k < n : ((\omega\times)^j)^k x_k \rangle\rangle\,(\mathsf{map}\,acc)\,y$

$\Longleftarrow$ { meaning of tri }

$\qquad \langle j : 0 \le j < n : \langle k : 0 \le k < n : x_k \rangle\rangle\,(\mathsf{tri}\,\mathsf{tri}(\times\omega)\,;\mathsf{map}\,acc)\,y$

$\Longleftarrow$ { meaning of *join* }

$\qquad \langle k : 0 \le k < n : x_k \rangle\,(J^{-1}\,;n\,;\mathsf{tri}\,\mathsf{tri}(\times\omega)\,;\mathsf{map}\,acc)\,y$

$\qquad$ where $J = apl^{-1}\,;\mathsf{rdl}\,join = apr^{-1}\,;\mathsf{rdr}\,join$

$\Longleftarrow$ $\quad x\,(J^{-1}\,;n\,;\mathsf{tri}\,\mathsf{tri}(\times\omega)\,;\mathsf{map}\,acc)\,y$

Since ω depends on n, because $\omega^n = 1$, we will be honest and write $(\times\omega)$ using a new operator $\otimes$ for which $z \otimes n = z \times \omega$. This operation has the property, which will be useful later, that $(\otimes(p \times q))^q = (\otimes p)$. There are two instances of reduction of an associative relation in this description, and we can avoid commitment about how to implement these by introducing a new operator red for which $\mathsf{red}\,R = apl^{-1}\,;\mathsf{rdl}\,R = apr^{-1}\,;\mathsf{rdr}\,R$ for any associative R.

The term in brackets that relates x to y represents the discrete Fourier transform algorithm, but only if x is a list of length n, so we will calculate

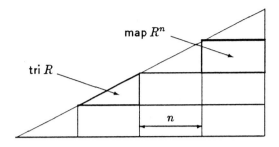

Figure 1.41: illustration of tri R divided into map n ; tri map R^n ; map tri R

from the definition

$$n \; ; \mathcal{F} \;\; = \;\; n \; ; (\text{red } join)^{-1} \; ; n \; ; \text{tri}\,\text{tri}(\otimes n) \; ; \text{map}\,\text{red } add$$

1.9.3 Dividing large problems into smaller ones

Suppose R is an algorithm or circuit for calculating some list-valued function of a list of values. If it is possible to express R in the form S ; red app, then S is an algorithm for constructing the same result in parts, and may be implementable by a number of independent parts. For example

$$\left(\text{red } join\right)^{-1} \;\; = \;\; \left(\text{red } join\right)^{-1} \; ; \text{map}\left(\text{red } join\right)^{-1} \; ; \text{red } app$$

describes a divide-and-conquer strategy for fanning out a signal into some number of copies, and then independently fanning out each of those.

Similarly red app ; R is an algorithm which constructs the same result as R from a partition of the same input into a list of lists. If it is possible to 'simplify' red app ; R into a form which has a parallel implementation, that gives a strategy for dividing the calculation of R. A particularly useful result in the present case is that

$$\text{map}\,n \; ; \text{red } app \; ; \text{tri } R \;\; = \;\; \text{tri}\,\text{map } R^n \; ; \text{map}\,\text{tri } R \; ; \text{map}\,n \; ; \text{red } app$$

which means that tri R can be implemented by a number of (smaller) independent instances of tri R and a triangular array of map R^n components. This equality depends on the restriction to a list-of-lists where every sublist has the same length.

In factorising the discrete Fourier transform, this rule is applied twice to an instance of an expression of the form tri tri R.

$$\text{map}\,\text{map}(\text{map } q \; ; \text{red } app) \; ; (\text{map } p \; ; \text{red } app) \; ; \text{tri}\,\text{tri } R$$
$$= \;\; \text{map}\,\text{map}(\text{map } q \; ; \text{red } app) \; ; \text{tri}\,\text{map } R^p \; ; \text{map}\,\text{tri } R \; ; \text{map}\,p \; ; \text{red } app$$

$$= \quad \text{tri map}(\text{tri map } R^{p \times q} \text{ ; map tri } R^p) \text{ ;}$$

$$\text{map map}(\text{map } q \text{ ; red } app) \text{ ; map tri } R \text{ ; map } p \text{ ; red } app$$

$$= \quad \text{tri map}(\text{tri map } R^{p \times q} \text{ ; map tri } R^p) \text{ ; map tri}(\text{tri map } R^q \text{ ; map tri } R) \text{ ;}$$

$$\text{map map}(\text{map } q \text{ ; red } app) \text{ ; map } p \text{ ; red } app$$

$$= \quad \text{tri map tri map } R^{p \times q} \text{ ;}$$

$$\text{tri map map tri } R^p \text{ ; map tri tri map } R^q \text{ ; map tri map tri } R \text{ ;}$$

$$\text{map map}(\text{map } q \text{ ; red } app) \text{ ; map } p \text{ ; red } app$$

This factorisation corresponds to the two changes of variables in the earlier calculation with summations. The order of the terms in R does not matter because terms in map and tri with the same relation necessarily commute with each other.

Since the R in question is $(\otimes n)$, this is the point to observe that some powers of R are going to be cancellable, specifically that

$$\text{tri map tri map}(\otimes n)^n \quad = \quad \text{map map map map}(\otimes 1)$$

$$= \quad \text{map map map map}(\times \omega)^0$$

where $(\times \omega)^0$ is the identity on the type underlying the arithmetic. That term can then be cancelled by absorbing it into any of the other three similar terms; this corresponds to the cancelling of ω^n in the calculations with summations.

1.9.4 Dividing the discrete Fourier transform

Suppose that $n = p \times q$. The factorisation of the n-point transform proceeds, as suggested above, by simplifying a specific instance of map q ; red app ; n ; $\mathcal{F}$. The particular instance is chosen – with hindsight, of course – so that a term can be cancelled later.

p ; map q ; red app ; $\mathcal{F}$

$$= \quad p \text{ ; map } q \text{ ; red } app \text{ ; } n \text{ ; } \mathcal{F}$$

$$= \{ \text{definition of } \mathcal{F} \}$$

$$p \text{ ; map } q \text{ ; red } app \text{ ; } (\text{red } join)^{-1} \text{ ; } n \text{ ; tri tri}(\otimes n) \text{ ; map red } add$$

$$= \{ \text{factorising red } join \}$$

$$p \text{ ; map } q \text{ ; red } app \text{ ; } (\text{red } join)^{-1} \text{ ; } (\text{map red } join)^{-1} \text{ ;}$$

$$q \text{ ; map } p \text{ ; red } app \text{ ; tri tri}(\otimes n) \text{ ; map red } add$$

$$= \{ \text{promoting red } app \text{ over joins} \}$$

$$p \text{ ; map } q \text{ ; } (\text{red } join)^{-1} \text{ ; } (\text{map red } join)^{-1} \text{ ; } q \text{ ; map } p \text{ ;}$$

$$\text{map map red } app \text{ ; red } app \text{ ; tri tri}(\otimes n) \text{ ; map red } add$$

$= \{$ factorising tri tri $\}$

  p ; map q ; (red $join$)$^{-1}$; (map red $join$)$^{-1}$;

  tri map map tri($\otimes n$)^p ; map tri tri map($\otimes n$)^q ; map tri map tri($\otimes n$) ;

  q ; map p ; map map red app ; red app ; map red add

$= \{$ promoting map red add over red app $\}$

  p ; map q ; (red $join$)$^{-1}$; (map red $join$)$^{-1}$;

  tri map map tri($\otimes q$) ; map tri tri map($\otimes p$) ; map tri map tri($\otimes n$) ;

  q ; map p ; map map(map red app ; red add) ; red app

$= \{$ associativity of add $\}$

  p ; map q ; (red $join$)$^{-1}$; (map red $join$)$^{-1}$;

  map tri tri map($\otimes p$) ; map tri map tri($\otimes n$) ; tri map map tri($\otimes q$) ;

  map map(map red add ; red add) ; q ; map p ; red app

This is a watershed in the calculation: the algorithm has now been teased apart into sufficiently many sufficiently small sub-calculations that we can begin to see how it might be re-arranged into smaller instances of the same algorithm.

The strategy from this point is to use a number of facts about the arithmetic to eliminate some map operators from the expression. Firstly the arithmetic is deterministic, and R ; $join^{-1} = join^{-1}$; $[R, R]$ for deterministic relations R, so (red $join$)$^{-1}$;map $R = R$;(red $join$)$^{-1}$. Secondly since multiplication distributes over addition, so does $\otimes$ and map($\otimes k$) ; red add = red add ; ($\otimes k$). To do this the order of some of the operators must be changed by composing both sides with a transposition.

trn ; p ; map q ; red app ; $\mathcal{F}$

  $= \{$ previous calculation $\}$

   q ; map p ; (red $join$)$^{-1}$; (map red $join$)$^{-1}$; map map trn ;

   map tri tri map($\otimes p$) ; map tri map tri($\otimes n$) ; tri map map tri($\otimes q$) ;

   map map(map red add ; red add) ; q ; map p ; red app

  $= \{$ transposing pointwise operations $\}$

   q ; map p ; (red $join$)$^{-1}$; (map red $join$)$^{-1}$;

   map tri map tri($\otimes p$) ; map tri tri map($\otimes n$) ; tri map tri map($\otimes q$) ;

   map map(trn ; map red add ; red add) ; q ; map p ; red app

  $= \{$ commutativity of add $\}$

   q ; map p ; (red $join$)$^{-1}$; (map red $join$)$^{-1}$;

   map tri map tri($\otimes p$) ; map tri tri map($\otimes n$) ; tri map tri map($\otimes q$) ;

   map map(map red add ; red add) ; q ; map p ; red app

$$= \{ \text{distributivity of } \otimes \text{ over } add \}$$
$$q \; ; \; \mathsf{map}\, p \; ; \; (\mathsf{red}\, join)^{-1} \; ; \; (\mathsf{map}\, \mathsf{red}\, join)^{-1} \; ; \; \mathsf{map}\, \mathsf{tri}\, \mathsf{map}\, \mathsf{tri}(\otimes p) \; ;$$
$$\mathsf{map}\, \mathsf{map}\, \mathsf{map}\, \mathsf{red}\, add \; ; \; \mathsf{map}\, \mathsf{tri}\, \mathsf{tri}(\otimes n) \; ; \; \mathsf{tri}\, \mathsf{map}\, \mathsf{tri}(\otimes q) \; ;$$
$$\mathsf{map}\, \mathsf{map}\, \mathsf{red}\, add \; ; \; q \; ; \; \mathsf{map}\, p \; ; \; \mathsf{red}\, app$$

$$= \{ \text{deterministic arithmetic} \}$$
$$q \; ; \; \mathsf{map}\, p \; ; \; (\mathsf{red}\, join)^{-1} \; ; \; \mathsf{tri}\, \mathsf{map}\, \mathsf{tri}(\otimes p) \; ;$$
$$\mathsf{map}\, \mathsf{map}\, \mathsf{red}\, add \; ; \; \mathsf{tri}\, \mathsf{tri}(\otimes n) \; ; \; (\mathsf{red}\, join)^{-1} \; ; \; \mathsf{tri}\, \mathsf{map}\, \mathsf{tri}(\otimes q) \; ;$$
$$\mathsf{map}\, \mathsf{map}\, \mathsf{red}\, add \; ; \; q \; ; \; \mathsf{map}\, p \; ; \; \mathsf{red}\, app$$

$$= \{ \text{promoting } q \text{ over joins} \}$$
$$\mathsf{map}\, p \; ; \; (\mathsf{red}\, join)^{-1} \; ; \; p \; ; \; \mathsf{tri}\, \mathsf{map}\, \mathsf{tri}(\otimes p) \; ; \; \mathsf{map}\, \mathsf{map}\, \mathsf{red}\, add \; ;$$
$$\mathsf{tri}\, \mathsf{tri}(\otimes n) \; ;$$
$$\mathsf{map}\, q \; ; \; (\mathsf{red}\, join)^{-1} \; ; \; q \; ; \; \mathsf{tri}\, \mathsf{map}\, \mathsf{tri}(\otimes q) \; ; \; \mathsf{map}\, \mathsf{map}\, \mathsf{red}\, add \; ;$$
$$\mathsf{red}\, app$$

There are two occurrences of similar expressions in the right-hand side, differing only in the parameter p or q. Each of these can be shown in the same way to satisfy

$$\mathsf{map}\, p \; ; \; (\mathsf{red}\, join)^{-1} \; ; \; p \; ; \; \mathsf{tri}\, \mathsf{map}\, \mathsf{tri}(\otimes p) \; ; \; \mathsf{map}\, \mathsf{map}\, \mathsf{red}\, add$$
$$= \quad \mathsf{map}\, p \; ; \; \mathsf{map}((\mathsf{red}\, join)^{-1} \; ; \; p) \; ; \; trn \; ; \; \mathsf{tri}\, \mathsf{map}\, \mathsf{tri}(\otimes p) \; ; \; \mathsf{map}\, \mathsf{map}\, \mathsf{red}\, add$$
$$= \quad \mathsf{map}(p \; ; \; (\mathsf{red}\, join)^{-1} \; ; \; p \; ; \; \mathsf{tri}\, \mathsf{tri}(\otimes p) \; ; \; \mathsf{map}\, \mathsf{red}\, add) \; ; \; trn$$
$$= \quad \mathsf{map}(p \; ; \; \mathcal{F}) \; ; \; trn$$

so showing that

$$q \; ; \; \mathsf{map}\, p \; ; \; trn \; ; \; \mathsf{red}\, app \; ; \; \mathcal{F}$$
$$= \quad \mathsf{map}(p \; ; \; \mathcal{F}) \; ; \; trn \; ; \; \mathsf{tri}\, \mathsf{tri}(\otimes n) \; ; \; \mathsf{map}(q \; ; \; \mathcal{F}) \; ; \; trn \; ; \; \mathsf{red}\, app$$

Now trn is its own inverse, and $p \; ; \; \mathsf{map}\, q \; ; \; \mathsf{red}\, app$ is also a bijection on the domain of the right-hand side, so both can be carried over to the other side of the equation.

$$n \; ; \; \mathcal{F} \quad = \quad (\mathsf{red}\, app)^{-1} \; ; \; trn \; ; \; \mathsf{map}(p \; ; \; \mathcal{F}) \; ; \; trn \; ; \; \mathsf{tri}\, \mathsf{tri}(\otimes n) \; ;$$
$$\mathsf{map}(q \; ; \; \mathcal{F}) \; ; \; trn \; ; \; \mathsf{red}\, app$$

The remaining asymmetry in the expression is annoying, but merely superficial for of course $trn \; ; \; \mathsf{tri}\, \mathsf{tri}(\otimes n) = \mathsf{tri}\, \mathsf{tri}(\otimes n) \; ; \; trn$.

So long as both p and q are strictly less than n we can use this decomposition as a definition of $n \; ; \; \mathcal{F}$ for any composite n. The decomposition can be read – taking terms from left to right – as a divide-and-conquer algorithm for

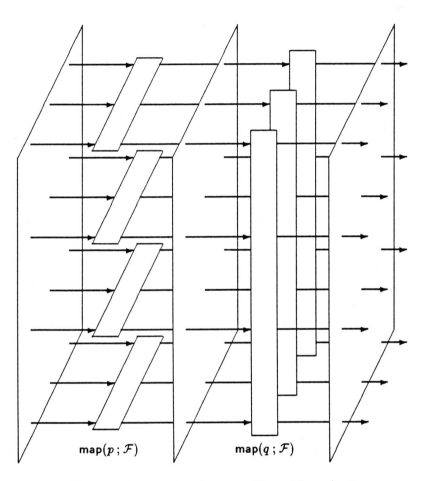

map(p ; $\mathcal{F}$)   map(q ; $\mathcal{F}$)

Figure 1.42: recursive decomposition of $(p \times q)$; $\mathcal{F}$

implementing transforms of composite width: divide the input into p chunks of length q; interleave them; apply an array (of q) independent p-point transforms; interleave the results; modify by scaling the $\langle i, j \rangle$-th signal by $(\otimes n)^{i \times j}$; apply an array (of p) independent q-point transforms; interleave the results; and finally concatenate the q resulting lists, each of which is of length p, into a single n-list. This is the algorithm known as the 'fast Fourier transform'. The scaling factors in $tri\,tri(\otimes n)$ can of course be pre-calculated for any given p and q and are known as 'twiddle factors'.

1.9.5   Outline of an implementation

The usual recursive 'butterfly' implementation of the fast Fourier transform applies only to transforms on vectors of length 2^n for some n. This is because it is very easy to do two-point transforms: because minus one is the principal square root of unity, the two-point transform $\Phi = 2\,;\,\mathcal{F}$ relates $\langle x_0, x_1 \rangle$ to $\langle x_0 + x_1, x_0 - x_1 \rangle$ and requires no multiplications.

For higher powers of two, the factorisation used is

$$2n\,;\,\mathcal{F} \quad = \quad pair\,;\,trn\,;\,\mathsf{map}(n\,;\,\mathcal{F})\,;\,trn\,;\,tri\,tri(\otimes 2n)\,;\,\mathsf{map}\,\Phi\,;\,trn\,;\,halve^{-1}$$

The only explicit multiplications in this factorisation are in the $tri\,tri(\otimes 2n)$ which can be implemented by an array of $2n$ multiplications only $n - 1$ of which are non-trivial. The factorisation is used recursively on the $n\,;\,\mathcal{F}$ term until only two-point transforms remain.

The usual way of implementing this algorithm – that is, the usual way of laying out the circuit – is to divide $\mathcal{F}$ into two homogeneous parts: let $\mathcal{F} = \mathcal{S}\,;\,\mathcal{F}'$ where

$$2n\,;\,\mathcal{F}' \quad = \quad halve\,;\,\mathsf{map}(n\,;\,\mathcal{F}')\,;\,tri\,tri(\otimes 2n)\,;\,((\mathsf{map}\,\Phi) \setminus trn)\,;\,halve^{-1}$$

and

$$2n\,;\,\mathcal{S} \quad = \quad pair\,;\,trn\,;\,\mathsf{map}(n\,;\,\mathcal{S})\,;\,halve^{-1}$$

Looking back to the discussion of butterfly circuits, you will see that the recursion

$$
\begin{aligned}
2n\,;\,\mathcal{B} \quad &= \quad (\mathsf{map}(n\,;\,\mathcal{B})\,;\,(\mathsf{map}\,\Phi) \setminus trn) \setminus halve^{-1} \\
&= \quad \mathsf{two}(n\,;\,\mathcal{B})\,;\,\mathsf{rpmap}\,\Phi
\end{aligned}
$$

has a solution $\mathcal{B} = (\bowtie \Phi^{-1})^{-1}$. This recursion is almost the same as that for $\mathcal{F}'$, and the solution need only be adjusted to accommodate the twiddle factors. Alternately, the twiddle factors might be calculated in a pre-pass using another co-located butterfly of the same shape as $\mathcal{B}$.

Similarly, the solution to the recursion for S is a permutation, which is a tree of transpositions. It is that very thorough shuffle which appears inscrutably in many implementations of the fast transform, and which reverses the bits of the index of the position of a value in a vector.

The derivation of the fast algorithm did not depend on a particular arithmetic: any integral domain would do. So you can use this derivation to lead into the design of a circuit in which the twiddle-factor multipliers operate on complex numbers, or on small integers with modulo arithmetic. More importantly, the chosen arithmetic can be the pointwise operations on time-sequences, and the derivation of the butterfly circuit can just as easily be used to lead to a sequential circuit.

1.10   References

[Bird88]   R. S. Bird, *Lectures on constructive functional programming*, Programming Research Group technical monograph PRG–69, September 1988.

[Cool65]   J. W. Cooley and J. W. Tukey, *An algorithm for the machine computation of complex Fourier series*, Mathematics of Computation, **19**, 1965. pp. 297–301.

[Dav89]   K. Davis and J. Hughes (eds.), *Functional Programming, Glasgow 1989*, Springer Workshops in Computing, 1990.

[Jon89]   G. Jones, *Deriving the fast Fourier algorithm by calculation*, in [Dav89]. (Programming Research Group technical report PRG–TR–4–89)

[Jon90]   G. Jones and M. Sheeran, *Relations and refinement in circuit design*, in [Morg90].

[Leis81]   C. E. Leiserson, *Area efficient VLSI computation*, Ph.D. thesis, Carnegie-Mellon University, Pittsburg, Pennsylvania, October 1981.

[McCab82] M. M. McCabe, A. P. H. McCabe, B. Abrambepola, I. N. Robinson and A. G. Cory, *New algorithms and architectures for VLSI*, GEC Journal of Science & Technology, Vol. 48, No. 2, 1982. pp. 68–75.

[Miln88]   G. J. Milne (ed.), *The fusion of hardware design and verification*, North-Holland, 1988.

[Morg90]   C. C. Morgan (ed.), *Proc BCS FACS workshop on Refinement*, Hursley, January 1990. (To appear from Springer 1990)

[Sheer88]   M. Sheeran,   *Retiming and slowdown in Ruby*,   in [Miln88]. pp. 289–308.

Formal Methods for VLSI Design   71
J. Staunstrup (Editor)
Elsevier Science Publishers B.V. (North-Holland)
© IFIP, 1990

Chapter 2

Synchronized Transitions

J. Staunstrup[1] and M.R. Greenstreet[2]

2.1 Introduction

Synchronized Transitions is a language for designing integrated circuits
by modeling them as parallel programs. Programs written in Synchronized
Transitions describe the computation as well as structure of digital circuits
and can be used to specify designs from very high levels of abstraction down
to gate level descriptions. Tools exist for simulation, formal verification, and
layout synthesis based on Synchronized Transitions descriptions of a design.

Synchronized Transitions programs are composed of **transitions** and
state variables. Transitions are guarded commands which model logic. State
variables model storage, and sharing of these variables models communication.
Non-deterministic execution of transitions corresponds to delays in hardware,
while their atomic semantics facilitates formal analysis and design verification.
Implementation conditions ensure that functional behavior is preserved
by hardware realizations.

Synchronized Transitions differs from many other guarded command
languages, e.g. CSP [9], because *shared variables* are used for communication
instead of messages. This provides a very simple correspondence between
programs and their hardware realizations.

---

[1] Department of Computer Science, Technical University of Denmark, DK-2800 Lyngby,
Denmark. e-mail: jst@iddth.dk
[2] Computer Science Department, Princeton University, 35 Olden St., Princeton, New
Jersey 08544-2087, USA. e-mail: mrg@princeton.edu

Synchronized Transitions also differs from typical hardware description languages, e.g., VHDL [8], Model [2], and ELLA [15], in that preconditions (guards) determine the temporal behavior. The language has no built-in clock concept or sequencing (";") constructs. There are strong similarities between Synchronized Transitions and UNITY, as developed by Chandy and Misra [1]. Both describe a computation as a collection of atomic conditional assignments without any explicit flow of control. Chandy and Misra propose this as a general programming paradigm. Our application of Synchronized Transitions is more specialized, in particular, the development of application specific VLSI circuits. The UNITY proof techniques may also be applied to Synchronized Transitions, and they provide a rich set of ideas for further research.

From the same high level description in Synchronized Transitions it is possible to synthesize either a synchronous or an asynchronous realization. This allows the designer to focus on issues of functionality rather than details of the realization. After completing the high level design, the designer can choose to employ either synchronous or asynchronous circuit techniques as is appropriate for the particular application. In either case, the designer verifies that the program meets certain implementation conditions which guarantee that the chosen circuit paradigm correctly implements the program semantics.

Synchronized Transitions is intended for designing application specific VLSI circuits. These are usually designed to satisfy requirements of size, maintainability, speed, or special function which cannot be met by off-the-shelf components. Synchronized Transitions facilitates experimenting with alternative architectures and offers the possibility of formal verification. Utilizing this potential reduces design costs by eliminating errors and wrong design decisions at an early stage of the design process.

2.2   Notation

VLSI circuits are composed of state variables, latches for storing these variables, and combinational logic for computing functions of state variables. Likewise, Synchronized Transitions programs are composed of state variables, and transitions for computing new values of these variables. Transitions specify state changes using multi-assignments. For example,

$$\ll \text{enable} \land \neg\text{reset} \rightarrow \text{out, s} := \text{in+s, in} \gg$$

is a transition which specifies an update of out and s. The new values of these state variables are in+s and in respectively. The value of s in in+s is the old value (before the state change). The update is only done if the precondition enable $\land\neg$reset holds. In general, a transition is of the form

$$\ll \textit{precondition} \rightarrow \textit{action} \gg$$

The **precondition** is a boolean expression. The **action** is a multi-assignment which specifies the state transformation made by the transition. It can only be performed if the precondition holds. In a multi-assignment, all expressions on the right hand side are evaluated and then assigned to the variables on the left hand side. All variables on the left hand side must be distinct. A transition is executed

repeatedly: each time it has been completed, it is immediately ready for a new execution,

atomically: it may be thought of as an indivisible operation,

independently: of the order it appears in the program text.

One may reason about a program by treating the transitions as a set of concurrent indivisible operations. As explained in sections 2.6.2 and 2.9.1, this abstract model allows efficient circuit realizations where many transitions are executed simultaneously.

It is not required that a transition be executed immediately after its precondition becomes satisfied; in fact, there is no upper bound on when it takes place. This is an abstraction of delays in hardware. For example,

$$\ll \text{ TRUE } \rightarrow \text{ y } := \text{a} \lor \text{b} \gg$$

describes an OR gate. The precondition, TRUE, specifies that it is always allowed to set the output, y, to the OR of the inputs, a and b; however, an arbitrary delay can elapse between a change of the inputs and the changing of the output.

A few special forms of transitions are used to simplify programs:

$\ll action \gg$
:   This is a transition whose precondition is always true. For example, $\ll$ y $:=$ a$\lor$b $\gg$ is equivalent to: $\ll$ TRUE $\rightarrow$ y $:=$ a$\lor$b $\gg$.

$\ll precondition \gg$
:   This is a transition which has a precondition, but performs no action. Such transitions are used in expressions formed with the combinators of the next section.

2.2.1   Instantiation and combinators

A transition describes the behavior of a subcircuit. Once such a subcircuit is fabricated, it is never removed, and it is continuously in operation (while the power is on). This behavior is modeled in a Synchronized Transitions program by a **transition instantiation** which yields a **transition instance** that is executed repeatedly, i.e., every time the precondition is satisfied, the action may be performed. To describe any significant circuit, a large number of such transition instances are needed. Three operators are provided for combining transition instances.

The **asynchronous combinator**, $\|$, combines two transition instances which execute independently. For example,

$$\ll \text{ a<b } \rightarrow \text{ a, b } := \text{ b, a } \gg \ \| \ \ll \text{ b<c } \rightarrow \text{ b, c } := \text{ c, b } \gg$$

specifies two independent transition instances, each of which may be executed whenever its precondition is satisfied. This computation sorts a, b, and c into descending order.

All transitions combined with $\|$ are independent; there is no global thread of control determining the order of execution. A program defines a set of transition instances. This set is static and completely determined by the program text. An operational model for the execution of a program in Synchronized Transitions is repeated selection and execution of an arbitrary transition instance from this set. No assumptions are made about the fairness of the selection mechanism.

The asynchronous combinator is the fundamental composition operator. Below two other combinators are introduced. They are convenient for structuring a program, but they can both be eliminated by rewriting the program into one using asynchronous combinators only.

The **synchronous combinator**, $+$, specifies that two transition instances are always executed together. For example,

$$\ll \text{ a } := \text{ b } \gg \ + \ \ll \text{ b } := \text{ d } \gg$$

is equivalent to

$$\ll \text{ a, b } := \text{ b, d } \gg$$

More generally,

$$\ll \text{ C}^1 \rightarrow \text{ a } := \text{ b } \gg \ + \ \ll \text{ C}^2 \rightarrow \text{ b } := \text{ d } \gg$$

is equivalent to

$$\ll \ \ \text{ C}^1 \wedge \ \ \text{ C}^2 \rightarrow \text{ a, b } := \text{ b, d } \gg \ \|$$
$$\ll \ \neg\text{C}^1 \wedge \ \ \text{ C}^2 \rightarrow \text{ b } := \text{ d } \gg \ \|$$
$$\ll \ \ \text{ C}^1 \wedge \neg\text{C}^2 \rightarrow \text{ a } := \text{ b } \gg$$

To avoid conflicting assignments, synchronous composition is only defined if the sets of variables written by the two transitions are disjoint. More formally, let W^i be the set of variables written by t^i and C^i be the precondition of t^i. The exclusive write condition is defined as follows:

Exclusive Write Condition: Two transition instances t^1 and t^2 meet the exclusive write condition if and only if:

$$(W^1 \cap W^2 \neq \emptyset) \Rightarrow \neg(C^1 \wedge C^2)$$

Intuitively, one may think of the execution of a group of synchronously composed transition instances as simultaneous execution of the entire group. This can be used to model a circuit controlled by a global clock, where all subcircuits (transition instances) operate once in every clock cycle.

It is possible to mix the two combinators + and $\|$ in one program. The meaning of such a program is defined by the rule given above for rewriting + into $\|$. However, most programs are constructed using only one of the combinators. For example + is typically used for synchronous designs and $\|$ for asynchronous ones.

The **product combinator**, $*$, specifies a merge of two transition instances. For example,

$$\ll \mathbf{C}^1 \rightarrow \mathbf{a} := \mathbf{b} \gg * \ll \mathbf{C}^2 \rightarrow \mathbf{b} := \mathbf{d} \gg$$

is equivalent to

$$\ll \mathbf{C}^1 \wedge \mathbf{C}^2 \rightarrow \mathbf{a}, \mathbf{b} := \mathbf{b}, \mathbf{d} \gg$$

The product combinator may only be applied to transitions meeting the exclusive write condition. One common use of the product combinator is to factor out a common sub-expression, e.g.,

$$\ll \neg\mathbf{reset} \gg * ($$
$$\ll \mathbf{C}^1 \rightarrow \ldots \gg \ \| \ \ll \mathbf{C}^2 \rightarrow \ldots \gg \ \| \ \ll \mathbf{C}^3 \rightarrow \ldots \gg \ )$$

This is equivalent to

$$\ll \mathbf{C}^1 \wedge \neg\mathbf{reset} \rightarrow \ldots \gg \ \| \ \ll \mathbf{C}^2 \wedge \neg\mathbf{reset} \rightarrow \ldots \gg \ \|$$
$$\ll \mathbf{C}^3 \wedge \neg\mathbf{reset} \rightarrow \ldots \gg$$

The three combinators have the following properties:

Commutative:	$t_1 \| t_2$	$\equiv$	$t_2 \| t_1$
	$t_1 * t_2$	$\equiv$	$t_2 * t_1$
	$t_1 + t_2$	$\equiv$	$t_2 + t_1$
Associative:	$(t_1 \| t_2) \| t_3$	$\equiv$	$t_1 \| (t_2 \| t_3)$
	$(t_1 + t_2) + t_3$	$\equiv$	$t_1 + (t_2 + t_3)$
	$(t_1 * t_2) * t_3$	$\equiv$	$t_1 * (t_2 * t_3)$
Precedence:	$t_1 \| t_2 * t_3$	$\equiv$	$t_1 \| (t_2 * t_3)$
	$t_1 \| t_2 + t_3$	$\equiv$	$t_1 \| (t_2 + t_3)$
	$t_1 + t_2 * t_3$	$\equiv$	$t_1 + (t_2 * t_3)$
Distributive:	$t_1 * (t_2 \| t_3)$	$\equiv$	$t_1 * t_2 \| t_1 * t_3$
	$t_1 + (t_2 \| t_3)$	$\equiv$	$t_1 + t_2 \| t_1 + t_3$

2.2.2   Transition declarations

Frequently, the same transition is required in many different places in a program. In each place, the transition instance references a different set of state variables. Synchronized Transitions provides a mechanism for parameterizing and naming transitions which is analogous to functions of traditional programming languages. For example,

```
TRANSITION copy(from, to: element);
≪ from≠E ∧ to=E → to, from := from, E ≫
```

Several instances of a transition can be created using the combinators described above and supplying actual parameters. For example,

```
copy(a, b) ∥ copy(x, y).
```

2.2.3   Cells

To make the design of large circuits tractable, it is necessary to have mechanisms for decomposing designs into independent parts that can be considered separately. This section presents the **cell** construct, which is used to modularize programs in Synchronized Transitions.

A cell encapsulates part of a design. It may contain local state variables, transitions, and subcells. The following cell defines the behavior of a subcircuit part that sets its local state variable local to FALSE.

```
CELL part;
  STATE local: BOOLEAN;
BEGIN
  ≪ local → local := false ≫
END part;
```

The internal structure of a cell is hidden from the rest of the circuit. The scope of state variables (and formal parameters) is defined to be the cell in which they are declared; the scope does not include inner subcells. Cells communicate via parameters to coordinate their computations. For example, the following cell communicates via the boolean parameter reset.

```
CELL part(reset: BOOLEAN);
  STATE local: BOOLEAN;
BEGIN
  ≪ reset → local := false ≫
END part;
```

Instantiation of a cell is specified by giving its name and an actual parameter list, e.g., part(init). An actual parameter may be read and written by both the instantiating and the instantiated cell unless access to the parameter is restricted by one of the mechanisms described in section 2.2.7.

A single state variable must not be known under two (or more) different names in a single cell. This is ensured by requiring that all actual parameters passed to a cell instance are distinct.

Cells may be nested to describe a hierarchy of subcircuits. For example,

```
CELL component(reset, in, out: BOOLEAN);
  STATE temp: BOOLEAN;
  CELL part(reset, o, t: BOOLEAN);
  BEGIN
    ≪ ¬reset → o := t ≫
  END part;
BEGIN
  part(reset, out, in) ‖ part(reset, temp, in)
END component;
```

describes a cell with an internal subcell that is instantiated twice with different actual parameters. In general, a cell defines the set of transition instances obtained by instantiating all elements in the body of the cell. Operationally, a program is a set of transition instances. This set is obtained by instantiating all cells, leading to instantiation of inner cells. This can lead to further inner instantiations, etc., until one reaches the innermost cells having transition instantiations only.

2.2.4 Static parameters

In addition to state variables, a cell may have static parameters. These are prefixed with the reserved symbol STATIC.

```
CELL component(reset: BOOLEAN; STATIC size: INTEGER);
```

Static parameters are used to control instantiations and to determine the size of data structures, e.g. arrays. The value of a static parameter must be fixed when the cell is instantiated; therefore, the corresponding actual parameter must be an expression consisting exclusively of constant values and static parameters.

2.2.5 Conditional instantiation and recursion

Conditional instantiation is used to control whether or not a particular subcell is instantiated. A conditional instantiation has the form

$$\{B \mid instantiation\}$$

where B is a boolean expression. It must be possible to determine the value of B statically; therefore, B may only contain static parameters and constant values, but no state variables.

Cells may be described recursively; i.e., a cell may instantiate a local copy of itself, for example,

```
CELL parity(in: word; STATIC low,high: bitrange; out: BOOLEAN);
  STATIC middle = (low+high) DIV 2;
  STATE lowparity, highparity: BOOLEAN;
  TRANSITION xor(a, b, y: BOOLEAN);
  ≪ y := a ≠ b ≫
  TRANSITION wire(a, y: BOOLEAN);
  ≪ y := a ≫
BEGIN
  { high-low ≥ 1 | ( parity(in, low, middle, lowparity)
      ‖ parity(in, middle+1, high, highparity)
      ‖ xor(lowparity, highparity, out) } ‖
  { high-low < 1 | wire(in[low], out) }
END parity;
```

Instantiating the cell parity(s, 0, 15, res) recursively instantiates 31 cells (16 wire cells and 15 xor cells). Note that the recursion unfolds during instantiation and *not while the transition instances execute*. Recursion is a description mechanism. The cell notation is used to structure the specification, and this structure controls the instantiation of transitions. Instantiation is *static*. Transition instances, like hardware, are neither dynamically created nor destroyed.

2.2.6   Quantified instantiation

Often it is necessary to create a number of similar instances of a transition or cell (operating on a set of state variables, for example, an array). This is expressed as follows:

{*combinator index : range | instantiation*}

The index is a name which can be used in the instantiation as a static parameter.

```
{ ‖ i: [0..n-1] | copy(a[i], a[i+1]) }
```

which specifies n instantiations, one for each value in the range [0..n-1]:

```
copy(a[0], a[1]) ‖ copy(a[1], a[2]) ‖ copy(a[2], a[3])
    ‖ ...‖ copy(a[n-1], a[n])
```

2.2.7 Restricting the use of state variables

A state variable passed as an actual parameter may be read and written both by the instantiating and by the instantiated cell. This unrestricted use can be restricted to prevent some state variables from being written in a particular scope. Three kinds of restrictions may be specified for a state variable or formal parameter (both of which are referred to as variables in the following).

- External write: the variable may be changed only externally, i.e., in a surrounding cell instance. This restriction is only meaningful for formal parameters.

- Local write: the variable may be changed in this scope only; it may not be changed in either internal or external cell instances.

- Internal write: the variable may be changed in internal cell instances only, i.e., by instances to which it is passed as an actual parameter.

These restrictions facilitate formal verification and optimization of circuits synthesized from programs in Synchronized Transitions. They are written among the declarations in a **restriction part**, e.g.,

```
RESTRICTION
    grp: EXTERNAL;
    reqp, grl, grr: LOCAL;
    reql, reqr: INTERNAL;
```

When a variable passed as an actual parameter is restricted to be externally or locally written, the corresponding formal parameter (in the cell instance to which the parameter is passed) must be declared to be external.

2.3 Invariants

An invariant is a predicate on the values of state variables. Once a program enters a state where this predicate is satisfied, all subsequent states will also satisfy the predicate. For example, an arbiter may not simultaneously grant two requests. If these grants are represented by the state variables grl and grr, this condition can be expressed by requiring that the boolean expression $\neg(grl \wedge grr)$ be invariantly true. Invariants can be stated in a **invariant part**, e.g.,

```
INVARIANT
    ¬(grl ∧ grr);
```

Invariants describe the *intended* operation of a program; they do not affect the execution of transition instances. Therefore, it is quite possible to write a program that does not maintain its stated invariants. One may show that the stated invariants are in fact maintained, either informally or by translating the program into a form which can be handled by a theorem proving program (see section 2.7.2). In this translation, invariants become proof obligations. In addition to providing an interface to a theorem prover, invariants could, in principle, be used by other translation programs. For example, a logic synthesis program could use invariants to produce a more efficient circuit.

2.4   Examples

Arbiter:
An *arbiter* is a circuit that provides indivisible access to a shared resource, e.g., a bus or a peripheral. The arbiter described here is implemented as a binary tree in which all nodes (including the root and the leaves) are identical. The arbitration algorithm is based on passing a unique token around the tree. An external process connected to a leaf may use the resource only when that leaf has the token.

Each node has three pairs of connections, one for its parent and one for each of its children. A connection pair consists of two signals, req and gr, standing for "request" and "grant." Such a pair is used according to the following four-phase protocol.

1. A node requests the token by setting req to TRUE.

2. When gr becomes TRUE, the node has the token and may pass it down the tree.

3. The token is handed back by setting req to FALSE.

4. When gr becomes FALSE, a new request can be made.

Figure 2.1 shows a few nodes and their interconnections. The signals reqp and grp at each node are connected to reql and grl, or to reqr and grr, at the next (higher) level of the tree (these connections are specified by the parameter passing). When both children of a particular node request the resource, it is given first to the left child; when that child releases the resource, it is given to the right child. A transition instance at the root copies reqp to grp so that the root is able to grant its own request immediately.

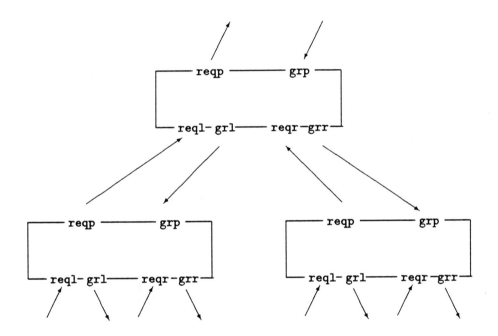

Figure 2.1: Two levels of arbiter tree

```
CELL arbiter(grp, reqp: BOOLEAN; STATIC level, max: INTEGER);
  STATE grl, reql, grr, reqr: BOOLEAN;
  RESTRICTION
    grp: EXTERNAL;
    reqp, grl, grr: LOCAL;
    reql, reqr: INTERNAL;
  INVARIANT
    ¬(grl ∧ grr) ∧ ( (grl ∨ grr ) => (grp ∧ reqp) );
BEGIN
  (* requestparent *)
  ≪ ¬grp ∧ (reql ∨ reqr) → reqp := TRUE ≫ ||
  (* grantleft *)
  ≪ grp ∧ reqp ∧ reql ∧ ¬grr → grl := TRUE ≫ ||
  (* grantright *)
  ≪ grp ∧ reqp ∧ reqr ∧ ¬grl ∧ ¬reql → grr := TRUE≫ ||
  (* doneleft *)
  ≪ ¬reql → grl := FALSE ≫ ||
  (* doneright *)
  ≪ ¬reqr → grr := FALSE ≫ ||
  (* done *)
  ≪ grp ∧ ¬grl ∧ ¬grr → reqp := FALSE ≫ ||
  (* instantiate children *)
  {level < max | arbiter(grl, reql, level+1, max) ||
                 arbiter(grr, reqr, level+1, max) }
END arbiter;
```

This arbiter is not fair. Furthermore, it has no external connections (at the leaves). Both of these shortcomings can be overcome. The simple version shown here is to illustrate the notation.

Traffic light:
A simple traffic light controller is used to illustrate the use of synchronous composition. Consider the scenario depicted in figure 2.2. A highway is crossed by a side road, and the intersection is controlled by a traffic light giving priority to traffic on the highway. At the intersection, there are sensors which can detect waiting cars. When a car on the side road has waited SideroadDelay time units, the light must change and allow the waiting car to cross. Similarly, a car waiting for HighwayDelay time units on the highway forces the light to change. The lights must change immediately to let a waiting car cross, if there are no cars waiting in the other direction.

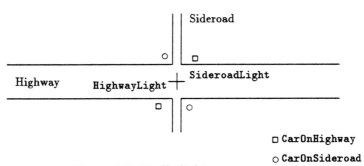

Figure 2.2: Traffic light

```
CELL trafficlight;
  STATIC Green = 0; Yellow = 1; Red = 2;
    SideroadDelay = 4; HighwayDelay = 1;
  TYPE color = [Green..Red];
CELL OneDirection(Reset, Car, CrossingCar: BOOLEAN;
              Light, OtherLight: color; STATIC limit: INTEGER);
  STATE timer: INTEGER;
BEGIN
  ≪ ((timer=0)∨¬Car)∧((Light=Green)∧CrossingCar)∧¬Reset
                    → Light:= Yellow ≫ +
  ≪ (timer>0)∧CrossingCar∧(Light=Green)∧¬Reset
                    → timer:= timer-1 ≫ +
  ≪ (Light=Yellow)∧¬Reset
       → Light, OtherLight, timer:= Red, Green, limit-1) ≫ +
  ≪ Reset → timer:= (limit - 1) ≫
END OneDirection;
STATE
  Reset: BOOLEAN; HighwayLight, SideroadLight: color;
  CarOnHighway, CarOnSideroad: BOOLEAN;
BEGIN
  OneDirection(Reset, CarOnHighway, CarOnSideroad,
              HighwayLight, SideroadLight, HighwayDelay) +
  OneDirection(Reset, CarOnSideroad, CarOnHighway,
              SideroadLight, HighwayLight, SideroadDelay) +
  ≪ Reset → HighwayLight, SideroadLight:= Green, Red ≫ +
  ≪ Reset → CarOnHighway, CarOnSideroad:= FALSE, FALSE ≫
END trafficlight;
```

2.5 Tools

We have several prototype tools supporting circuit design using Synchronized Transitions. These tools are translators that transform source programs in Synchronized Transitions into various other forms. There are three main classes of such translators, helping the designer at different phases of the design:

synthesis: produces circuit descriptions, e.g., netlists or layouts,

verification: produces verification conditions which can be checked by a mechanical theorem prover,

simulation: produces an executable program, for example in C, that can be used for informal verification.

The ultimate goal of a VLSI design is the construction of an efficient circuit (layout). However, even the fastest and most compact design is useful only if it is correct and implements the intended function. Therefore, the constructs of a VLSI design language should facilitate verification. In the case of Synchronized Transitions, formal and automated verification has been given top priority in the design of the language. It is currently possible to use a translator which transforms a program into proof obligations which are then processed by a mechanical theorem prover [5]. Exactly the *same* source program can be fed to the synthesis program which transforms it into a layout. This is an application of the What You Prove Is What You Fabricate (WYPIWYF) principle. This prevents errors form creeping into a design when it is manually transformed from one form oriented towards synthesis (traditionally, a circuit diagram) to another form oriented towards verification (typically, a symbolic expression). Formal verification is discussed further in section 2.7.

2.6 Synchronous realizations

The previous sections have shown how the functional behavior of a circuit may be specified using Synchronized Transitions. This facilitates design at a high level of abstraction which is very important for evaluating capabilities and trade-offs. In this section, it is shown how simple and efficient synchronous circuits can be derived from a program in Synchronized Transitions.

2.6.1 Two-phase realizations

Consider a transition $t = \ll C^t \rightarrow l^t := F(r^t) \gg$. Figure 2.3 sketches a direct synchronous realization of t. $\boxed{C^t}$ is a combinational network which

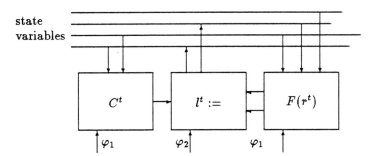

Figure 2.3: Two-phase realization

computes the precondition, $\boxed{F(r^t)}$ is another combinational network which computes the values of the expressions of the transition's multiassignment, and $\boxed{l^t}$ implements the storage of these state variables. State variables may be realized in many different ways, e.g. with flip/flops or, in MOS circuits, using the capacitance of wires. Realizations of $\boxed{C^t}$ and $\boxed{F(r^t)}$ can be derived using logic synthesis programs, the details are not discussed here. Note that *each transition instance can be synthesized independently* of the others.

Synchronous composition

A program with many transition instances is realized by synthesizing a subcircuit like the one in figure 2.3 for each instance. If all transitions in the program are composed with the synchronous combinator $(+)$, then the entire program is realized by directly connecting the subcircuits, an example is shown in figure 2.4. The placement of transition instances and the corresponding wiring has profound influence on the area needed to realize a program. In general, by carefully structuring the Synchronized Transitions program, the designer can express a locality of communication that leads to efficient realizations.

Asynchronous composition

In this section, we describe how programs using the asynchronous combinator can be realized. Each transition instance is synthesized exactly as it was described in figure 2.3. In general, it is not possible to combine transition instances realized this way and preserve the behavior of a program using asynchronously composed transitions. This is because simultaneous execution of several enabled transition instances may not be equivalent to atomic execution (one transition instance at a time). Consider the following example:

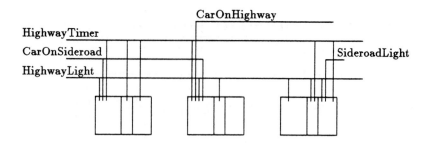

Figure 2.4: Direct connection of two-phase realizations

```
CELL example;
  STATE a, b: BOOLEAN;
  INVARIANT a ∨ b;
  TRANSITION set(p,q: BOOLEAN);
  ≪ TRUE → p:= ¬q ≫
BEGIN
  set(a,b) ∥ set(b,a)
END example;
```

Simultaneously execution of the two transition instances may lead to states which do not satisfy the invariant. For example, if both a and b are initially true, simultaneous execution of both transition instances will lead to a state with both a and b false, violating the invariant of the original program. Hence, a two-phase realization does not always preserve the atomicity of the transitions.

There are two ways of solving this problem. One would be to find another realization which performs the transition instances in the example according to the program. Another approach is to exclude programs such as the one shown above. We take the second approach and restrict our programs to ones where the direct combination of two-phase realizations as shown in figure 2.4, works. The restrictions on a program are formulated as a set of *implementation conditions* which are sufficient to ensure correct operation of the realization.

2.6.2   Implementation condition CREW

In this section, we characterize a class of programs with asynchronously composed transitions for which the two-phase realization is correct. In the two-phase realization, *all* enabled transition instances are executed in *every* clock cycle. This is a correct way of implementing a Synchronized Transitions program with asynchronously composed transitions where dependent transition instances are never enabled simultaneously. Two transition instances are **dependent** if one writes a state variable which the other reads or writes. This implementation condition can be defined more precisely as follows: let $R^i = \{r_1, r_2, \ldots\}$ be the set of variables read by a transition instance t^i and $W^i = \{w_1, w_2, \ldots\}$ be the set of variables written by t^i. C^i is the precondition of t^i.

Concurrent Read Exclusive Write Condition (CREW):
For all distinct pairs of transition instances t^1 and t^2:
$$(R^1 \cap W^2 \neq \emptyset) \vee (R^2 \cap W^1 \neq \emptyset) \vee (W^1 \cap W^2 \neq \emptyset) \Rightarrow \neg(C^1 \wedge C^2)$$

The CREW condition says, that it must not be possible to satisfy the precondition of a transition instance while another dependent transition instance is enabled. Establishing $\neg(C^1 \wedge C^2)$ may require more analysis than just a local argument using boolean algebra. Typically, an invariant capturing some global property of the program is needed. It is important to note *that this condition is verified by checking the* Synchronized Transitions *program*, and the checking can be done at an early phase of the design.

To be a sound implementation condition, CREW must ensure that the states which can be reached by the realization can also be reached by atomic execution of transition instances from the original program. To see this, consider one full clock cycle (both phases) starting in a state S and leading to a state S'.

Let $(s_1, s_2, \ldots, s_m)$ be the state variables written in this clock cycle. Because CREW is satisfied, each is written by exactly one transition instance, and none of them are read by any of the other transition instances $(t^1, t^2, \ldots, t^{i-1}, t^{i+1}, \ldots, t^n)$. Hence, writing $(s_1, s_2, \ldots, s_m)$ (by the same transition instances) in any serial order leads to state S'.

$$S \qquad\qquad\qquad\qquad\qquad\qquad\qquad\qquad\qquad S'$$

$$t^1 :\ll \quad \ldots \quad \gg$$

$$t^2 :\ll \ldots \gg$$

$$t^n :\ll \ldots \gg$$

Any state which is obtained by a two-phase realization of a program satisfying CREW can also be obtained by executing the transition instances one at a time (atomically). Therefore, the two-phase realization is sound in the sense that, any invariant of the original program is also maintained by the two-phase realization. CREW is a *sufficient* condition for two-phase realizations. However, CREW is not necessary; there are (weaker) conditions that suffice. Weaker conditions are more complicated and unnecessary for the examples in these notes.

Transformations to satisfy CREW

The CREW condition is very strong, and many programs do not satisfy it. However, there are several transformations which can be used to obtain a program satisfying CREW. Consider two transitions not satisfying CREW:

$$
\begin{aligned}
t^1 &= \ll C^1 \rightarrow A^1 \gg \\
t^2 &= \ll C^2 \rightarrow A^2 \gg
\end{aligned}
$$

As t^1 and t^2 are dependent, the conjunction, $C^1 \wedge C^2$, may be satisfied. To satisfy CREW, C^1 and C^2 can be modified to guarantee that their conjunction is always false. A decision must be made about which transition instance to execute when both of the original preconditions are satisfied. Let the boolean expression, B, represent the outcome of this decision. If B is true, the first transition instance is executed, otherwise the second. The two transition instances are therefore rewritten as follows:

$$
\begin{aligned}
u^1 &= \ll B \vee \neg C^2 \gg * t^1 &= \ll C^1 \wedge (B \vee \neg C^2) \rightarrow A^1 \gg \\
u^2 &= \ll \neg B \vee \neg C^1 \gg * t^2 &= \ll C^2 \wedge (\neg B \vee \neg C^1) \rightarrow A^2 \gg
\end{aligned}
$$

The preconditions of u^1 and u^2 exclude each other. Regarding B, there are several alternatives. The simplest is to make a static choice about which transition instance should be executed, e.g., if B is the constant TRUE:

$$
\begin{aligned}
u^1 &= t^1 &= \ll C^1 \rightarrow A^1 \gg \\
u^2 &= \ll \neg C^1 \gg * t^2 &= \ll (C^2 \wedge \neg C^1) \rightarrow A^2 \gg
\end{aligned}
$$

If this is not adequate, the choice can be made to alternate by letting B be a boolean variable b assigned in the two transitions as follows:

$$u^1 \;=\; \ll C^1 \wedge (\; \text{b} \vee \neg C^2) \rightarrow A^1 \gg \; * \; \ll \text{b} := \text{FALSE} \gg$$
$$u^2 \;=\; \ll C^2 \wedge (\neg \text{b} \vee \neg C^1) \rightarrow A^2 \gg \; * \; \ll \text{b} := \text{TRUE} \gg$$

Finally, b could also be implemented by circuitry producing a random value; in which case, the transition instance to execute is chosen non-deterministically.

2.7 Mechanical verification

One of the major advantages of using a high level notation for circuit design is the potential for thorough verification in the early phases of the design process. By verifying each description of the design before proceeding to the next level of detail, the designer can avoid wasting time implementing detailed circuits that would later be discarded. This is advantageous, since it is expensive to locate and correct errors in the later stages of a project; for example, if correcting these errors requires extensive, global changes to the design. Furthermore, the long turn-around time for circuit fabrication makes it attractive to use techniques which uncover errors at an early phase of the design.

There are many types of verification. For example, the designer may wish to show that the circuit computes the correct function, that it operates at the desired speed, that its power consumption is acceptable, or that it can be sold profitably. In this chapter, we consider the formal verification of functional properties. By formal verification, we mean rigorous proofs that the design has a precisely stated property. In particular, we consider functional properties that can be expressed as invariants. The simple semantics of Synchronized Transitions is well suited for this approach. In section 2.7.1, we describe how invariants may be used to formulate correctness criteria for VLSI designs and give several examples.

Any nontrivial VLSI design has many details. The formal verification must consider these details. Accordingly, a manually constructed proof is as tedious and subject to error as the original design. Fortunately, by using mechanical theorem provers, this complexity can be made manageable. In section 2.7.2, we describe our experiences using one such theorem prover, the "Larch Prover" - LP [3, 4].

2.7.1 Invariants

An invariant is an assertion involving state variables which must hold between the execution of transition instances [11]. Let I be an invariant, $S.pre$ the state of program immediately before some action is taken, and $S.post$ the state after the action. To show that I is an invariant, we show that if $I(S.pre)$ is satisfied, then $I(S.post)$ must also hold. Once an invariant holds, it continues to hold. See section 2.7.2 for a discussion of initialization.

If all transitions in a program are asynchronously composed (using $\|$), then actions correspond to transition instances. To show that I is an invariant, it is sufficient to consider each transition t in the program and show that executing an instance of t preserves the invariant. In particular, for $t = \ll C^t \rightarrow A^t \gg$, one must show that if $I \wedge C^t$ holds, then executing A^t will lead to a state in which I continues to hold. Typically, the difficult part of formal verification is finding an appropriate invariant I. Once the correct invariant is formulated, verifying it is often straightforward (although often tedious, which is why mechanical theorem provers are helpful).

In these notes, we only consider programs where all transitions have been asynchronously composed. This is the fundamental way of writing a program in Synchronized Transitions. In principle, all other combinators can be eliminated using the properties given in section 2.2.1.

Examples

1. The traffic light was discussed in section 2.4. If HighwayLight=Green or HighwayLight=Yellow, cars may enter the intersection from the highway. Likewise, for the side road light and cars. To prevent collisions, these two situations must not happen at the same time. This can be expressed with the invariant

   $$(\text{HighwayLight} = \text{Red}) \vee (\text{SideroadLight} = \text{Red})$$

2. Consider the arbiter described in section 2.4. We assume that the interface between the arbiter and each of the external clients is a pair of signals, req and gr, standing for "request" and "grant." There is a separate pair (req[i], gr[i]) for each of its n clients. The key property of the arbiter is that it never grants access to two different clients simultaneously. This is expressed with the invariant

   $$i \neq j \Rightarrow \neg(\text{gr}[i] \wedge \text{gr}[j])$$

3. We return to the traffic light to formulate an additional invariant expressing that the waiting time of a car is limited. To express this a slightly richer language for writing invariants is needed. This language is obtained by augmenting the original program with auxiliary variables [18]. These are state variables which are used in the verification only; they are not needed to execute the program. In principle, the auxiliary variables could be declared and used like other state variables. This

requires additional "auxiliary" transitions. For the traffic light, we use auxiliary variables to record old values of state variables. In particular, x.pre denotes the value of x *before* the last transition instance was executed, so x < x.pre expresses that executing the last transition instance decreased x. For example, with the transition ≪ x:= x-1 ≫, the manipulation of the auxiliary variable x.pre could be written as follows:

$$≪ \; \text{x:= x-1} \; ≫ \; * \; ≪ \; \text{x.pre:= x} \; ≫$$

Similarly, an auxiliary variable x.post denotes the value of x after executing the transition instance. The auxiliary transitions are usually omitted to avoid cluttering the original program with extra details. Instead, we use the convention that x.pre denotes the previous value of x and x.post the current value. Should there be doubt about the meaning of this, one may always resolve it by constructing the augmented program with auxiliary variables where each update of x is accompanied by an update of x.pre recording the "old" value of x.

The following assertion expresses the limited waiting time at the traffic light:

$$(\text{timer.post} \geq 0) \wedge (\text{timer.pre} > \text{timer.post})$$

This is, however, too strong to be an invariant, because some transition instances do not change timer. It is, for example, changed only when cars are waiting on the side road. Instead, we suggest the following invariant:

$$((\text{CarOnSideroad.post} \wedge \text{HighwayLight.post} = \text{Green})) \; \Rightarrow$$
$$(\text{timer.post} \geq 0) \wedge (\text{timer.pre} > \text{timer.post})$$

This expresses that the car on the side road only waits a limited number of steps before the light on the highway changes from green to red. With auxiliary variables, a rich variety of properties of Synchronized Transitions programs can be formulated as invariants.

Verification of an invariant

In this section a simplified version of the traffic light is used to illustrate how an invariant is verified. In the simplified traffic light all transitions are composed with the asynchronous combinator ||.

```
 CELL trafficlight;
    STATIC Green = 0; Yellow = 1; Red = 2;
    STATE
      Reset: BOOLEAN;
      HighwayLight, SideroadLight: INTEGER;
      CarOnHighway, CarOnSideroad: BOOLEAN;
    INVARIANT (HighwayLight=Red) ∨ (SideroadLight=Red);
  BEGIN
    (* ------- Change from Green to Red on highway --------- *)
    ≪ (HighwayLight = Green) ∧ CarOnSideroad ∧ ¬ Reset
                 → HighwayLight:= Yellow ≫ ||
    ≪ (HighwayLight = Yellow) ∧ ¬ Reset
                 → HighwayLight,SideroadLight:= Red,Green ≫ ||

    (* ------ Change from Green to Red on sideroad ------- *)
    ≪ (SideroadLight = Green) ∧ CarOnHighway ∧ ¬ Reset
                 → SideroadLight:= Yellow ≫ ||
    ≪ (SideroadLight = Yellow) ∧ ¬ Reset
                 → SideroadLight,HighwayLight:= Red,Green ≫ ||

    (* ------ Initialization -------------------------- *)
    ≪ Reset → HighwayLight, SideroadLight, CarOnHighway,
                 CarOnSideroad:= Green, Red, FALSE, FALSE ≫
  END trafficlight;
```

The first transition is

```
≪ (HighwayLight = Green) ∧ CarOnSideroad ∧ ¬ Reset
                 → HighwayLight:= Yellow ≫
```

This transition cannot violate the invariant, because it changes HighwayLight from Green to Yellow; therefore, SideroadLight must have been Red both before and after the transition was executed. The second transition is

```
≪ (HighwayLight = Yellow) ∧ ¬ Reset
                 → HighwayLight,SideroadLight:= Red,Green ≫
```

This cannot violate the invariant because HighwayLight is made Red. The two transitions for the side road light are symmetrical. Finally, the transition

```
≪ Reset → HighwayLight, SideroadLight, CarOnHighway,
                 CarOnSideroad:= Green, Red, FALSE, FALSE ≫
```

makes SideroadLight Red; again the invariant is maintained.

We have shown how one invariant of the traffic light can be verified by considering each transition of the program. A complete verification of a program requires proving a large collection of such properties, and typical programs have many more than the five transitions of the traffic light. For such applications, manual proofs, such as the one above, become very tedious and error prone. By automating much of the verification process, theorem proving programs, such as LP, make the verification process much more practical.

2.7.2 LP: the Larch Prover

LP is a theorem prover, based on equational term-rewriting, for a fragment of first-order logic [3, 4]. It has been used to analyze formal specifications written in Larch [7], to reason about algorithms involving concurrency, and to establish the correctness of hardware designs [5, 20]. The intended applications of LP motivate several departures in features and design from other term-rewriting programs and theorem provers.

- LP is most often used to debug a specification or a set of invariants. Hence, it is more important to report when and why a proof breaks down than to automatically explore many avenues for pushing a proof through to a successful conclusion. For this reason, LP does not employ heuristics to derive subgoals mechanically from conjectures. Instead, it relies largely on forward rather than backward inference, with the user rather than the program being responsible for inventing useful lemmas.

- LP is designed to work with large sets of large equations. It has been used to verify circuit designs described by several hundred equations.

- LP permits users to define theories by means of equations, induction schemes, and other (nonequational) rules of deduction.

- LP provides facilities that allow users to establish subgoals and lemmas during the proof of a theorem. It supports a variety of proof methods beyond those found in conventional equational term-rewriting. In the intended applications of LP, rewriting techniques such as normalization, completion [10], and proof by consistency [3] are often inapplicable, awkward, or computationally too expensive. Other techniques, such as proofs by cases or induction, often lead to better results.

Using LP with Synchronized Transitions programs

Synchronized Transitions programs are composed of state variables and transitions. To produce a LP script, state variables are translated to variables, and transitions are translated to equations. A variable is introduced with a "declare" statement, for example,

```
declare Reset:-> BoolId
```

Equations state relationships between variables. They are introduced with an "add" statement, for example

```
add Green = 0
```

defines Green to be the integer constant 0.

Transitions can be described as equations relating the states before and after execution. Using the .pre and .post conventions as described in section 2.7.1, we can translate the transition

```
≪ Reset → HighwayLight, SideroadLight, CarOnHighway,
            CarOnSideroad:= Green, Red, FALSE, FALSE ≫
```

into the LP construct

```
when t(pre, post) yield
  ((Reset).pre)
  (((HighwayLight).post) = Green)
  (((SideroadLight).post) = Red)
  (((CarOnHighway).post) <=> FALSE)
  (((CarOnSideroad).post) <=> FALSE)
```

This equation describes the relationship between the .pre and .post values of the state variables. To make the description of the transition complete, conjuncts must be added to the equation stating that the values of all variables which are not written by the transition are the same before and after executing the transition. This is done automatically by the translator described in the next section.

An asynchronous composition of transitions specifies a non-deterministic selection of the transition instance to be executed. In LP, one writes the disjunction of the corresponding equations. For example

$$t_1 \parallel t_2 \parallel t_3 \ldots \parallel t_n$$

becomes

```
add P(pre, post) = t₁(pre, post) ∨ t₂(pre, post) ∨
                    t₃(pre, post) ... tₙ(pre, post)
```

For P to be true, the equation for one of $t_1 \ldots t_n$ must be satisfied. The equation that is satisfied corresponds to the transition instance that was (non-deterministically) selected for execution.

To verify an invariant, a proof goal is stated. For example

```
prove I(pre) ∧ P(pre, post) ⇒ I(post)
```

Typically, the proof in LP is carried out by case analysis on the transitions in P. This corresponds to considering the transitions one at a time, as was done in the manual proof in section 2.7.1.

Automatic translation

The usage of LP is facilitated by using a translator which forms a complete LP proof script from a program in Synchronized Transitions. The translator performs the transformations described above. It also adds clauses for the state variables that are unchanged by the transition. This is an easily automated bookkeeping task. In LP, various idioms are used to guide the rewriting process. For example, the equations x=y, x->y, and x<=>y all express the equality of x and y; however, each construction activates different rewrite rules. Likewise, the assertion x => y and the deduction rule "when x yield y" both express implication but are handled differently in the rewriting process. For each construction in Synchronized Transitions, the translator produces the idiom that we have found to be the most practical. As a result, LP often verifies invariants of Synchronized Transitions programs directly from the output of the translator without any manual assistance.

As an example, the translator transforms the simple traffic light program into the following proof script (this is the actual output of the translator, with some explanatory comments added):

```
%% first execute a script to define basic concepts
%% (boolean algebra, some properties of integers, etc.)
exe prefix
%% define the colors Green, Yellow, and Red
declare Green:-> Int; add Green -> 0
declare Yellow:-> Int; add Yellow -> 1
declare Red:-> Int; add Red -> 2

%% declare state variables
declare Reset:-> BoolId
declare HighwayLight:-> IntId; declare SideroadLight:-> IntId
declare CarOnHighway:-> BoolId; declare CarOnSideroad:-> BoolId

%% declare invariant
set name invariant
declare Inv: State -> Bool
add
Inv(ss) ==
    (((HighwayLight).ss) = Red) | (((SideroadLight).ss) = Red)
..
%% define the transitions
set name transition
declare
    t1, t2, t3, t4, t5, actionTaken:  State, State -> Bool
```

```
..
add-immune-deduction-rules
  when actionTaken(pre, post) yield
    Inv(pre) &
    ( t1(pre, post) | t2(pre, post) | t3(pre, post) |
      t4(pre, post) | t5(pre, post) )

  when t1(pre, post) yield
    (((((HighwayLight).pre) = Green) & ((CarOnSideroad).pre)) &
    (not(((Reset).pre))))
    (((HighwayLight).post) = Yellow)
    ((((((Reset).post) <=> ((Reset).pre)) & (((SideroadLight
    ).post) = ((SideroadLight).pre))) & (((CarOnHighway).post)
    <=> ((CarOnHighway).pre))) & (((CarOnSideroad).post) <=>
    ((CarOnSideroad).pre)))

  when t2(pre, post) yield
    ((((HighwayLight).pre) = Yellow) & (not(((Reset).pre))))
    (((HighwayLight).post) = Red)
    (((SideroadLight).post) = Green)
    (((((Reset).post) <=> ((Reset).pre)) & (((CarOnHighway
    ).post) <=> ((CarOnHighway).pre))) & (((CarOnSideroad).post)
     <=> ((CarOnSideroad).pre)))

  when t3(pre, post) yield
    (((((SideroadLight).pre) = Green) & ((CarOnHighway).pre)) &
    (not(((Reset).pre))))
    (((SideroadLight).post) = Yellow)
    ((((((Reset).post) <=> ((Reset).pre)) & (((HighwayLight
    ).post) = ((HighwayLight).pre))) & (((CarOnHighway).post)
    <=> ((CarOnHighway).pre))) & (((CarOnSideroad).post) <=>
    ((CarOnSideroad).pre)))

  when t4(pre, post) yield
    ((((SideroadLight).pre) = Yellow) & (not(((Reset).pre))))
    (((SideroadLight).post) = Red)
    (((HighwayLight).post) = Green)
    (((((Reset).post) <=> ((Reset).pre)) & (((CarOnHighway
    ).post) <=> ((CarOnHighway).pre))) & (((CarOnSideroad).post)
     <=> ((CarOnSideroad).pre)))
```

```
when t5(pre, post) yield
  ((Reset).pre)
  (((HighwayLight).post) = Green)
  (((SideroadLight).post) = Red)
  (((CarOnHighway).post) <=> FALSE)
  (((CarOnSideroad).post) <=> FALSE)
  (((Reset).post) <=> ((Reset).pre))
..
%% now prove the invariant
set name theorem
prove actionTaken(pre, post) => Inv(post) by case
   actionTaken(pre, post)
  resume by cases
  t1(pre, post) t2(pre, post) t3(pre, post) t4(pre, post)
  t5(pre, post)
..
```

Experience with LP

We have used LP for the verification of several designs. Our experience from
these experiments is described in [5, 20] and summarized below:

- Circuit verification seems more amenable to machine based checking
  than traditional program verification.

- Combined with Synchronized Transitions, the technique of invariant
  assertions is useful for machine based reasoning about circuits.

- Even for simple circuits, one cannot rely on arguments that are not
  machine based.

- The additional effort required to produce a machine based proof is primarily for replacing vague, and therefore suspect arguments of an informal argument with sound, mechanizable methods.

- With the current level of theorem prover technology some experience is
  required to construct the guidance needed in machine based proofs.

- Machine based proofs are easily repeatable. This makes it easy to check
  minor adjustments of a design.

Formal hardware verification is a new field in which we see much potential.
Currently, the techniques and tools for formal verification are not sufficiently
refined for wide spread application. In spite of these limitations, we have
found formal verification to be helpful in gaining a better understanding of

VLSI design and in uncovering mistakes in some designs that were not found in simulation. We expect that, with further research and experience, formal verification will become of increasing importance.

Initialization

The techniques described in the previous sections demonstrate that a program (set of transition instances) preserves an invariant. When this is the case, we know that once the program enters a state satisfying the invariant, it remains in such states. The initialization must ensure that the program enters a state satisfying it.

In **Synchronized Transitions**, there are no built-in initialization constructs. From a formal verification viewpoint they would be convenient; however, such constructs would conceal the correspondence between programs and circuits. A circuit must be initialized explicitly, e.g., by giving a reset signal. Likewise, the initialization in a **Synchronized Transitions** program must be specified explicitly. If all initializations are collected in one transition, for example,

```
TRANSITION init;
 ≪ reset → v₁, v₂, ...:= e₁, e₂, ...≫
```

proper initialization can be proven by showing $init(pre, post) \Rightarrow I(post)$. We do not have an adequate verification technique for formally showing proper initialization when the initialization involves several transitions.

Hierarchical verification

Given a description of a circuit as a hierarchical composition of cells, we would like to reason about its behavior in an analogous hierarchical fashion. Once we have proven the design of a cell to be correct, we would like to be able to use as many instances of the cell as we please without being forced to reconsider the cell's correctness argument for each instance. In general, this is not possible. Typically, the correct operation of a cell depends on the its formal parameters (i.e. state variables shared with other cells) being, in some sense, well-behaved. For example, to show that CREW is maintained, we cannot permit unrestricted writing of an output of a cell by transition instances outside the cell.

We use **protocols** to specify what changes may be made to variables shared between cells, for example, that external transition instances may not modify certain variables, or that the client of an arbiter may not withdraw an ungranted request. Protocols separate the proofs of what happens inside a cell from what is done by the cell's environment. With an arbiter protocols may be used to show that the arbiter only grants one request at a time as long as the requests obey a certain protocol. Once this is proven, instances of the arbiter may be used in many different contexts. It is not necessary to reprove the correctness of the arbiter cell in each context, it is only necessary to show that each context maintains the protocol. Details of this approach are given in [20].

Beyond safety properties

In the preceding sections, we have shown how invariants can be verified and how the verification can be supported by mechanical tools. Similar techniques and machine support are available for showing that a program meets implementation conditions such as CREW (see section 2.6.2) or the conditions for delay insensitive circuits (see section 2.9.1). In the next section, it is shown how one may formally verify that one program implements another, using abstraction functions. These are all safety properties; they prevent "bad things from happening," for example, entering states not satisfying the invariant. As shown in section 2.7.1, use of auxiliary variables enables us to formulate and show a kind of progress. For example, properties of the form "either the program does nothing or the following will happen ..."

Much research is currently directed towards more powerful formal verification techniques for demonstrating liveness properties, i.e., to show that certain events *must* occur. We have chosen *not* to consider formal liveness proofs for many reasons. The most important is to simplify the verification procedure. Formal and mechanized verification of safety properties is already rather difficult. In fact it is so difficult that more work is needed to demonstrate its applicability beyond the academic exercises such as the ones shown in these notes. We are convinced that formal proofs of circuits will be of practical use only if they are supported by mechanical tools such as a theorem prover. When using such tools, it must always be kept in mind that we are up against undecidable properties and combinatorial explosions. This forces us to be humble and not pursue impossible goals.

One may then ask: "what is the point of verifying some safety properties, if it cannot be ensured that the final circuit will perform even a single step of the specified computation?" We offer two answers to this. First, there are many desirable properties that we do not verify formally, for example, power consumption, timing, and noise sensitivity. All these are verified separately with other techniques, mostly informal. Presently, we are leaving liveness among these. Secondly, the safety proof ensures that if a circuit ever delivers an output, then it meets the verified properties. This is a significant guarantee. In practice, we believe that most design errors manifest themselves by circuits giving the wrong answer rather than by not giving any answer at all.

2.8   Abstraction functions

This section describes a formal technique for showing that one design in
Synchronized Transitions, called the concrete program is a correct implementation of another design, called the abstract program. To be a correct
implementation, the concrete program must be closely related to the abstract
one, yet there must be significant differences between the two (otherwise, one
of them is superfluous). This relationship can be captured by an **abstraction function**, $\mathcal{R}$, mapping states of the concrete program to states of the
abstract.

_____ abstract design _____

$\mathcal{R}$

_____ concrete design _____

Often, it is not possible to go directly from the abstract design to the final
concrete circuit design. Instead one goes through a series of intermediate
designs where each new (lower) level is more concrete than the previous. The
following is a typical series of refinements:

_____ functional level _____

$\mathcal{R}_3$

_____ gate level   _____

$\mathcal{R}_2$

_____ transistor level _____

$\mathcal{R}_1$

_____ layout   _____

Between each of these levels, there is an abstraction function relating the
lower level (more concrete) to the one above (more abstract).

Formally, an abstraction function is a mapping from the concrete state
space, S_c, to the abstract, S_a:

$$\mathcal{R} : S_c \mapsto S_a$$

S_c is the set of all possible states, i.e., all the combinations of values that
the concrete state variables may take. Similarly, S_a is the set of all possible
abstract states.

In this treatment, we only consider programs where all transition instances
are composed using the asynchronous combinator ($\|$).

Example: A Modulo-4 counter

Consider a simple modulo-4 counter. It can be described abstractly as follows:

```
STATE
    incrA: BOOLEAN; cA: INTEGER;
BEGIN
    ≪ incrA → incrA, cA:= FALSE, (cA+1) MOD 4 ≫
END
```

The signal `incrA` indicates that an increment should be made. When it has been done `incrA` becomes false again.

Such a counter can be realized with two bits (`1C`, `mC`) representing `cA`. `1C` is the least significant bit and `mC` the most significant in the binary representation of `cA`.

```
STATE
    incrC: BOOLEAN; 1C, mC: BOOLEAN;
BEGIN
    ≪incrC∧ ¬1C → incrC, 1C:= FALSE, TRUE ≫ ||
    ≪incrC∧ 1C∧mC → incrC, 1C, mC:= FALSE, FALSE, FALSE ≫ ||
    ≪incrC∧ 1C∧ ¬mC → incrC, 1C, mC:= FALSE, FALSE, TRUE ≫
END
```

A more succinct description of the concrete counter combines the three transitions into the following one (where $\oplus$ denotes the exclusive OR operation):

```
    ≪ incrC → incrC, 1C, mC:= FALSE, ¬1C, 1C⊕mC ≫
```

To illustrate how abstraction functions can be used with programs of many transitions, we will show that the first concrete program (with three transitions) correctly implements the abstract program. It is straightforward to adopt this proof to the second, more succinct program.

The abstraction function must relate the concrete state x_c consisting of `incrC` and (`1C`, `mC`) to the abstract state (`incrA`, `cA`). Formally, this is specified as follows:

$$\mathcal{R}(x_c) = make_a(2 * I(\text{mC}.x_c) + I(\text{1C}.x_c), \text{incrC}.x_c)$$

Here, $make_a$: INTEGER, BOOLEAN $\mapsto S_a$ is a function which transforms a pair (INTEGER, BOOLEAN) to an abstract state[3]. The indicator function I : BOOLEAN $\mapsto$ INTEGER converts a boolean to an integer ($I(\text{TRUE}) = 1$ and $I(\text{FALSE}) = 0$).

---

[3] The function $make_a$ maps each of its arguments to a component of the abstract state. In particular, $\text{cA}.make_a(\text{n},\text{b}) = \text{n}$, and $\text{incrA}.make_a(\text{n},\text{b}) = \text{b}$.

2.8.1   Conditions on the abstraction function

In this section, we define conditions on an abstraction function which ensure that a concrete program does only what the abstract program prescribes. However, since we are concerned with safety properties only, the condition does not require that the concrete program does everything that is possible in the abstract one.

Executing transition instances of the concrete program must preserve the invariants of the abstract program. This is guaranteed if for every execution of a concrete transition instance t_c, there is an abstract transition instance, t_a, whose execution would have the equivalent effect. This is formalized by the following requirement:

$$\forall t_c : \Big(\{pre_c\} \ t_c \ \{post_c\} \ \Rightarrow$$

$$\exists t_a : \ \{\mathcal{R}(pre_c)\} \ t_a \ \{\mathcal{R}(post_c)\}\Big)$$

This requirement captures the central idea of relating the behavior of a concrete implementation to its abstract specification: for each action in the concrete program, there is a corresponding action (an "explanation") in the abstract program. There are, however, two major problems with the above requirement.

The first problem is that this requirement is, in a sense, too strong. Typically, a concrete program will include details that are not in the abstract program. These details may include state variables that are not present in the abstract program, for example the intermediate carry bits in an adder, or the internal state of a finite state machine. If a transition instance only changes these variables, there is no corresponding change to the abstract state. That is, $\mathcal{R}(pre_c) = \mathcal{R}(post_c)$ when such a transition instance is executed, but the condition stated above requires the existence of a corresponding abstract transition. This problem is discussed further in section 2.8.5.

The second problem with the above requirement is that it is, in another sense, too weak. If $\mathcal{R}$ maps all concrete states to a constant, the concrete program appears (in the abstract state space) to do nothing. For a disabled transition instance, t_a, we have $\{\mathcal{R}(pre_c)\} \ t_a \ \{\mathcal{R}(post_c)\}$, so the concrete program vacuously implements the specification. For example, if incrA.$\mathcal{R}(S_c)$ = FALSE and cA.$\mathcal{R}(S_c)$ = 0 for all concrete states S_c, the abstract program is trivially implemented *no matter what the concrete program does!* In the next section, this problem is addressed in the context of abstraction functions for cells.

2.8.2   Cells

When cells are used to structure a design, it is not sufficient to require that the states of the concrete design correspond to the states of the abstract: we also want the concrete cell to present an interface to its environment that is consistent with the interface of the abstract cell. The interaction between a cell and its environment is through the formal parameters of the cell. Since, the environment of the cell cannot see the cell's internal state variables, the interpretation of the formal parameters should be independent of the values of the cell's internal state variables. This leads to a restriction on $\mathcal{R}$. Let F_c and L_c denote the formal parameters and local variables respectively of a concrete cell. Let F_a and L_a denote the same for the abstract cell. We require that the value of $F_a.\mathcal{R}(F_c, L_c)$ be independent of L_c. That is,

$$\forall f, l, l' : \; F_a.\mathcal{R}(f, l) = F_a.\mathcal{R}(f, l')$$

An abstraction function that meets this condition is called **well formed**. With a well formed abstraction function, the values of the formal parameters of the abstract cell depend only on the formal parameters of the concrete one. We write $\mathcal{R}_f$ to denote this restriction of the abstraction function to formal parameters. That is,

$$F_a = \mathcal{R}_f(F_c)$$

The well formedness condition says that the values of the abstract formal parameters can only change when the concrete parameters do. However, the converse is not enforced by the well formedness condition; in particular, the formal parameters of the concrete cell can change without a corresponding change to the parameters of the abstract cell. As mentioned in the previous section, this allows trivial implementations. To prevent this, we require that a change in the formal parameters of the concrete cell map to a change in the formal parameters of the abstract cell. This ensures that if the concrete cell does anything (visible to its environment), a similar change is observed in the abstract program. More formally, we require $\mathcal{R}_f$ to be one-to-one,

$$\forall f, f' : \; (\mathcal{R}_f(f) = \mathcal{R}_f(f')) \Leftrightarrow (f = f')$$

An abstraction function that meets this condition is called **interface preserving**. Any interface preserving abstraction function is also well formed.

Example: A Modulo-4 counter (continued)

In the modulo-4 counter, there are three transition instances in the concrete program:
Consider the first one, assuming $pre_c, post_c$ are such that:

$$\{pre_c\} \; \texttt{incrC} \wedge \neg \texttt{1C} \rightarrow \texttt{incrC}, \texttt{1C} := \texttt{FALSE}, \texttt{TRUE} \; \{post_c\}$$

From this, it can be concluded that:

$$\mathbf{cA}.\mathcal{R}(pre_c) \;=\; 2 * I(\mathrm{mC}.pre_c) + I(\mathrm{1C}.pre_c) = 2 * I(\mathrm{mC}.pre_c) \leq 2$$

$$
\begin{aligned}
\mathbf{cA}.\mathcal{R}(post_c) &= 2 * I(\mathrm{mC}.post_c) + I(\mathrm{1C}.post_c) \\
&= 2 * I(\mathrm{mC}.post_c) + 1 \\
&= 2 * I(\mathrm{mC}.pre_c) + 1 \\
&= \mathbf{cA}.\mathcal{R}(pre_c) + 1 \\
&= (\mathbf{cA}.\mathcal{R}(pre_c) + 1) \ \mathtt{MOD} \ 4
\end{aligned}
$$

$$
\begin{aligned}
\mathrm{incrA}.\mathcal{R}(pre_c) &= \mathtt{TRUE} \\
\mathrm{incrA}.\mathcal{R}(post_c) &= \mathtt{FALSE}
\end{aligned}
$$

Hence, it follows that

$$\{\mathcal{R}(\mathrm{prec})\} \ \mathrm{incrA} \rightarrow \mathrm{incrA}, \mathrm{cA} := \mathtt{FALSE}, (\mathrm{cA} + 1) \ \mathtt{MOD} \ 4 \ \{\mathcal{R}(\mathrm{postc})\}$$

The reasoning for the other two transition instances is similar.

2.8.3   Utilizing LP

Section 2.7.2 described, how the theorem prover LP could be used to verify invariants. The proof shown above can be automated in a similar way. The abstract transition is described as follows:

```
trA(preA, postA) ==
  (incrA.preA) & ( (incrA.postA) = false ) &
  ( (cA.postA) = next(cA.preA) )
```

The function next is defined elsewhere as modulo 4 increment. The three concrete transitions are described as follows:

```
tr1C(preC, postC) ==
  ( (incrC.preC) & not(1C.preC)) &
  ( ((incrC.postC) = false) & ((1C.postC) = true) ) &
  ( (mC.postC) = (mC.preC) )
tr2C(preC, postC) ==
  ( (incrC.preC) & (1C.preC) & not(mC.preC)) &
  ( ((incrC.postC) = false) & ((1C.postC) = false)  &
  ( (mC.postC) = true) )
tr3C(preC, postC) ==
  ( (incrC.preC) & (1C.preC) & (mC.preC)) &
  ( ((incrC.postC) = false) & ((1C.postC) = false)  &
  ( (mC.postC) = false) )
```

The abstraction function uses an auxiliary function makeA to construct an abstract state out of its components. This is defined as follows:

```
(cA.makeA(i1, b1)) == i1
(incrA.makeA(i1, b1)) == b1
makeA( (cA.x_a), (incrA.x_a) ) == x_a
(makeA(i1, b1) = makeA(i2, b2)) == (i1 = i2) & (b1 <=> b2)
```

Finally, the abstraction function itself is described in this way:

```
I(true) == 1  I(false) == 0
R(x_c) -> makeA( (2*I(mC.x_c)) + I(1C.x_c), (incrC.x_c) )
```

Using these definitions, the abstraction condition is verified by showing an implication of the following form, for each of the three concrete transition instances:

```
prove tr1C(preC, postC) => trA(R(preC), R(postC)) )
```

2.8.4 Inheriting invariants

In this section, it is shown how an invariant of an abstract program can be inherited by a concrete implementation. An invariant guarantees that a program will only enter some subset of the possible state space. By inheriting an invariant from the abstract program, it is only necessary to define the abstraction function for states that are allowed by the invariant (i.e., it can be a partial function). Since impossible states do not have to be considered, both the formulation of the abstraction function and the verification that the implementation is correct can be simplified. Consider again two programs related by an abstraction function, $\mathcal{R}$.

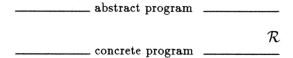

Assume that I is an invariant for the abstract program, and that $\mathcal{R}$ meets the requirement stated in section 2.8.1, then there is a corresponding invariant in the concrete program, this is called $\mathcal{R}^{-1}(I)$. There is no guarantee that there is a simple expression for $\mathcal{R}^{-1}(I)$. To see whether a concrete state x_c satisfies $\mathcal{R}^{-1}(I)$ one must check and see if $\mathcal{R}(x_c)$ satisfies I. The "dot" notation may also be used with invariants, so $I.x_a$ means I is satisfied in the state x_a, similarly $\mathcal{R}^{-1}(I).x_c$ means that the concrete invariant is satisfied in state x_c (i.e. $\mathcal{R}^{-1}(I).x_c \equiv I.\mathcal{R}(x_c).$)

To show that $\mathcal{R}^{-1}(I)$ is an invariant for the concrete program, assume that $\mathcal{R}^{-1}(I).pre_c$ and that

$$\{pre_c\}\ t_c\ \{post_c\}$$

Then it follows that:

$$\exists t_a : \{\mathcal{R}(pre_c)\}\ t_a\ \{\mathcal{R}(post_c)\}$$

Since I is an invariant for the abstract program, $I.\mathcal{R}(post_c)$ must hold, and hence, $\mathcal{R}^{-1}(I).post_c$. Therefore, $\mathcal{R}^{-1}(I)$ is an invariant for the concrete program.

When I is an invariant, $\mathcal{R}^{-1}(I)$ defines a subset of the concrete state space which contains all states that will occur. To show that a function $\mathcal{R}$ meets the abstraction condition, it is therefore only necessary to consider state pairs $pre_c, post_c$ such that

$$\{pre_c \wedge \mathcal{R}^{-1}(I).pre_c\}\ t_c\ \{post_c\}$$

Furthermore, the concrete program may have an invariant restricting the concrete states that can occur, call this invariant I_c. I_c may use concrete variables not appearing in the abstract program.

By utilizing the invariants, it is possible to give a less restrictive requirement on the abstraction function

$$\forall t_c : \Big(\{pre_c \wedge \mathcal{R}^{-1}(I).pre_c \wedge I_c\}\ t_c\ \{post_c\}\ \Rightarrow$$

$$\exists t_a : \{\mathcal{R}(pre_c)\}\ t_a\ \{\mathcal{R}(post_c)\}\Big)$$

This ensures that for every execution of a concrete transition instance there is a corresponding abstract transition instance. This condition is still too strong, in the next section we discuss how to weaken it.

2.8.5   The abstraction condition

Usually, the concrete program contains more details than the abstract. Therefore, the abstraction function maps many concrete states to the same abstract state. Such redundancy can be necessary to achieve the desired efficiency or robustness in the implementation. An example is the representation of bits used in section 2.9 to realize circuits which are independent of delays in wires and gates. Often in a redundant implementation, the concrete program makes state changes between two different concrete states representing the same abstract state $(\mathcal{R}(preC) = \mathcal{R}(postC))$. Such changes are called invisible, because they are not reflected in the abstract program. The requirement on the abstraction function formulated above does not allow for this, so a weaker condition is needed. The one presented in the previous section requires that there is an abstract transition instance corresponding to *every* concrete one. It is, however, only necessary to require this for every transition instance making a visible state change.

These considerations are captured by the **abstraction condition**

$$\forall t_c : \Big(\{pre_c \wedge \mathcal{R}^{-1}(I).pre_c \wedge I_c\} \ t_c \ \{post_c\} \ \Rightarrow$$

$$(\exists t_a : \{\mathcal{R}(pre_c)\} \ t_a \ \{\mathcal{R}(post_c)\}) \ \vee \ (\mathcal{R}(pre_c) = \mathcal{R}(post_c)) \Big)$$

If we have two programs, an abstract and a concrete, and an interface preserving abstraction function meeting the abstraction condition then all changes on the interface of the concrete program are reflected by changes on the interface of the abstract. Since the abstraction function is interface preserving, there is a functional relationship between the formal parameters of the concrete program and the abstract. Furthermore, whenever the formal parameters of the concrete program change (then the second disjunct of the abstraction condition cannot be satisfied), the abstract programs is forced to make a similar change. This is ensured by the first clause of the disjunct in the abstraction condition.

Definition 1 *A program P_c implements another program P_a iff there exists an interface preserving abstraction function, $\mathcal{R}$, meeting the abstraction condition.*

Example: A FIFO

A FIFO queue is a data structure which holds a sequence of elements. There are two external operations on a FIFO queue: insertion and removal. Elements are always removed in the same order as they are inserted (**First In First Out**). An element may be inserted in the queue when the input register is empty (in.s = FALSE). Insertion is done by assigning an element to in. An element may be removed when the output register is full (out.s = TRUE). An abstract FIFO can be described as follows:

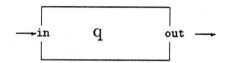

```
TYPE node = RECORD s: BOOLEAN; b: element; END;
CELL fifoA(inA, outA: node);
   STATE qA: SEQUENCE OF element;
BEGIN
   ≪ inA.s → qA, inA.s:= qA⊙inA.b, FALSE ≫ ||
   ≪ ¬outA.s ∧ qA=e⊙r → qA, outA.s, outA.b:= r, TRUE, e ≫
END
```

This description uses the type SEQUENCE, which is assumed to have a concatenation operator ⊙. In the final circuit, the sequence may be realized with a sequence of shift registers. However, first we consider an intermediate representation using an array of registers.

```
CELL fifoC(inC, outC: node);
   STATE qC: ARRAY[0..n-1] OF node;
TRANSITION transfer(to, from: node): BOOLEAN;
≪ from.s AND NOT to.s → from.s, to:= FALSE, from ≫
BEGIN
   transfer(qC[0], inC) || transfer(outC, qC[n-1]) ||
   {|| i: [0..n-2] | transfer(qC[i+1], qC[i]) }
END
```

The FIFO is represented as an array of registers qC[i]. The contents of one register is transferred to the next by an assignment, this corresponds to a bit parallel realization. The abstraction function is[4]:

$$\mathcal{R}(x_c) \quad = \quad make_a(inC.x_c, outC.x_c, F(qC.x_c, n-1))$$

where

$$F(q, i) \quad = \quad \begin{array}{ll} \lambda & i < 0 \\ F(q, i-1) & q[i].s = \text{FALSE} \\ q[i].b \odot F(q, i-1) & q[i].s = \text{TRUE} \end{array}$$

λ is the empty sequence. This abstraction function is interface preserving since inC is mapped to inA and outC is mapped to outA. It must also be shown that $\mathcal{R}$ meets the abstraction condition. There are three cases (three kinds of concrete transition instances) to consider.

---

[4] As in the example with the modulo-4 counter, the function $make_a$ maps each of its arguments to a component of the abstract state. In particular, $inA.make_a(inC, outC, x) = inC$, $outA.make_a(inC, outC, x) = outC$, and $qA.make_a(inC, outC, x) = x$.

1. $\{pre_c\}$ `transfer(qC[0], inC)` $\{post_c\}$
   We know that:
   `inC.s.`pre_c = `TRUE`, `qC[0].s.`pre_c = `FALSE`, `inC.s.`$post_c$= `FALSE`,
   `qC[0].s.`$post_c$ = `TRUE`, and `qC[0].b.`$post_c$ = `inC.b.`pre_c.

$$
\begin{aligned}
\text{qA}.\mathcal{R}(post_c) &= F(\text{qC}.post_c, n-1) \\
&= F(\text{qC}.pre_c, n-1) \odot \text{qC}[0].b.post_c \\
&= \text{qA}.\mathcal{R}(pre_c) \odot \text{qC}[0].b.post_c \\
&= \text{qA}.\mathcal{R}(pre_c) \odot \text{inC}.b.pre_c
\end{aligned}
$$

$$
\begin{aligned}
\text{inA}.s.\mathcal{R}(post_c) &= \text{FALSE} \\
\text{inA}.s.\mathcal{R}(pre_c) &= \text{TRUE} \\
\text{inA}.n.\mathcal{R}(pre_c) &= \text{inC}.b.pre_c
\end{aligned}
$$

Hence, it follows that:
$\{\mathcal{R}(pre_c)\}$ `inA.s` $\rightarrow$ `qA, inA.s:=` `qA`$\odot$`inA.b, FALSE` $\{\mathcal{R}(post_c)\}$

2. $\{pre_c\}$ `transfer(qC[i+1], qC[i])` $\{post_c\}$
   We know that: `qC[i].s.`pre_c = `TRUE`, `qC[i+1].s.`pre_c = `FALSE`.

$$
\begin{aligned}
\text{qA}.\mathcal{R}(post_c) &= F(\text{qC}.post_c, n-1) \\
&= F(\text{qC}.pre_c, n-1) \\
&= \text{qA}.\mathcal{R}(pre_c)
\end{aligned}
$$

This concrete transition instance makes an invisible state change. A syntactical check ensures that none of the formal parameters (`inC`, `outC`) are changed.

3. The third transition instance (output) is similar to the first.

From these three cases, it follows that $\mathcal{R}$ meets the abstraction condition.

Further restrictions on the abstraction function

The abstraction condition allows for very significant differences between the abstract and the concrete programs. Consider, for example, the FIFO where the abstract program allows for an unbounded number of elements in the FIFO. In contrast, the concrete program has a fixed capacity of n elements. In some cases this latitude is desirable, but there are cases where further restrictions on the abstraction function are needed. One could, for example, require that the abstraction function is surjective (every abstract state is represented by at least one concrete state). Such a requirement would not allow a concrete program with fixed capacity to represent an abstract one with unbounded capacity.

We have only considered programs where all transition instances are composed with ||. Although all other programs can be rewritten to this form, it is not practical to do so. It would, therefore, be useful to formulate a different abstraction condition for programs composed with +.

2.9   Delay insensitive realizations

Section 2.6 presented a technique for implementing a synchronous realization of a program. In this section, another type of realization is described; it leads to a delay insensitive circuit. Delay insensitive circuits operate correctly regardless of time delays caused by wires and logic elements. Such circuits can operate as fast as the technology, implementation, operating conditions, and specific input data allow.

Delay insensitive circuits form a special case of asynchronous circuits. In asynchronous designs, no global clock is used. Instead, the progress of computation is controlled by the values of state variables. Asynchronous circuits can be classified according to the assumptions that are made about the realization. Regardless of which category is considered, the criteria for correct operation is the same: *the functionality of the (atomic) program must be preserved by the asynchronous realization.* In particular, all invariants must be preserved. We divide asynchronous circuits into three categories.

- **Self-timed:** Upper and lower bounds may be specified for the delays of circuit elements and wires. These delays are exploited to produce a working design.

- **Speed-independent:** Circuit elements are allowed to have arbitrary, positive, finite delays, but wires are assumed to have no delays.

- **Delay insensitive:** Both circuit elements and wires may have arbitrary, positive, finite delays.

These classes are distinguished by the types of assumptions that are allowed about the timing behavior of the realization. Self-timed designs admit the largest class of assumptions, and delay insensitive admit the least. Any circuit that is speed-independent is also self-timed. Likewise, any circuit that is delay insensitive is speed-independent and therefore self-timed. Since delay insensitive designs must be verifiable with the fewest assumptions, they comprise the smallest of the three classes of circuits. On the other hand, this lack of timing assumptions enables us to give a simple formal characterization of delay insensitivity; accordingly, they are well suited for formal verification. We present such a formal characterization in the following paragraphs.

The semantics of `Synchronized Transitions` allows active transition instances to be executed in any order. This non-determinism appears as an arbitrary delay between satisfying the precondition of a transition and the subsequent execution. The non-determinism corresponds to the gate delays in the circuits which realize the transition.

In section 2.3, the state variables of a `Synchronized Transitions` program were associated with signals of the physical circuit. When a state variable in a `Synchronized Transitions` program is modified, the new value is immediately available to all transitions which read the variable. However, the changes of a physical signal between logical values may not occur at the same time for all circuit elements. This incoherence arises from many causes, for example, the time for a change of voltage to be transmitted along a wire, or variations in threshold voltages that cause two elements to switch at different points of the change. In a synchronous design, the clock period is chosen to be long enough to guarantee that all signals have settled before the values stored in latches are updated. For a delay insensitive design, it must be shown that the implementation functions correctly in the presence of these effects. These effects can be modeled as arbitrary delays (see [6]).

Let x be a state variable of a program. To model an arbitrary delay, we introduce a new state variable, x' and the **wire transition**

$$\ll x' := x \gg$$

The non-determinism in the execution of this transition models an arbitrary delay. Let t be a transition that reads x. To model an arbitrary delay between a change of x and the detection of this change by t, we modify t so that it reads x' instead of x. By introducing a new transition and state variable for each transition that reads x, changes of x can arrive at these transitions in any order.

We will now give a more rigorous definition of delay insensitivity. First, we formalize the notion of modifying a transition to read a delayed version of a state variable. Consider the transformation of a program P to a program P' with one wire transition. Let S be the set of state variables of P, T be the transition instances of P, x be a variable in S, and t be a transition instance in T that reads x. The program with the x input of t arbitrarily delayed, P', has state variables S' and transition instances T' such that

$$T' = T - \{t\} \cup \{t'\} \cup \{\ll x' := x \gg\}$$
$$S' = S \times x'$$

Where t' is obtained from t by replacing all occurrences of x with x'. If P' can be obtained from P by a transformation of this form, then we say that P' is a **single-delayed version** of P. We say P_n is a **delayed version** of P_1 if there exists a finite sequence of programs, $P_1, \dots P_n$ such that for each

$i : 1 < i \leq n$, P_i is a single-delayed version of P_{i-1}. Note that this definition allows cascaded wire transitions for the same input of a transition.

Definition 2 *A program P is* **delay insensitive** *if and only if every delayed version of P, implements P.*

Let $P"$ denote an arbitrary delayed version of P. To prove that a program is delay insensitive, we can use induction on the sequence of transformations from P to $P"$. For the base step, we note that for sequences of length one, $P" = P$, and $P"$ implements P trivially. For the induction step, we want a property R, such that for any programs Q and Q' where Q' is a single delayed version of Q:

$$R.1 : \quad R(Q) \Rightarrow Q' \text{ implements } Q$$
$$R.2 : \quad R(Q) \Rightarrow R(Q')$$

If we can find such a property, then $R(P)$ implies that P is delay insensitive. Because R is applied to the *original* program P, it is never necessary to consider programs with explicit wire delays. We have found this approach to be much less tedious than methods where programs with explicit wire delays must be considered.

In section 2.9.1, we propose two implementation conditions which together constitute a property R that guarantees delay insensitivity. In section 2.9.2, we prove that these conditions have properties $R.1$ and $R.2$. We then show some applications and limitations of this approach.

2.9.1   Implementation conditions

Consumed values

Consider the AND gate shown in figure 2.5. This circuit is not delay insensitive. Without wire delays, if the a input is pulsed true, and later b is pulsed, the y output will remain false. However, if the a input is delayed by a wire transition, the delayed version, a', may be true when b is pulsed. This could cause the y output to become true (a "glitch"). Hence, the realization can exhibit a different behavior than the original program. The consumed values condition addresses problems of this form. Informally, the consumed values condition requires all transition instances that read a particular state variable to do so between any two consecutive changes of the value of the variable. This ensures that old values of the variable are not pending in a wire when the value of the variable is changed.

To define the consumed values condition formally, a few auxiliary concepts are needed.

- The read set, R^y, of a state variable, y, is the set of all transition instances reading y.

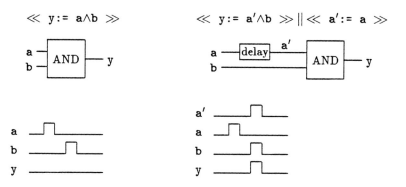

Figure 2.5: An AND gate with and without delay

- A transition instance is **active** if its precondition is satisfied, and executing the instance would change the value of one or more variables that it writes.

- t **depends** on the state variable y, if t is active and t would not be active if y had a different value.

Note that a transition instance can be enabled without being active if none of the variables it writes change by executing the action. For each state variable, y, we introduce a set valued auxiliary variable, a^y. The set a^y contains those transition instances which have executed depending on y since the last time y was changed. Each time y is changed, a^y becomes empty, and every time a transition instance, t, that depends on the variable y is executed, t is added to a^y. Initially, $a^y = R^y$. These auxiliary variables are used in the definitions only; they are not included in the final circuit.

Definition 3 *A program P satisfies* **consumed values** *if and only if for each state variable, y, $a^y = R^y$ when the value of y is changed.*

The consumed values condition requires all transition instances in R^y to be active and executed between any two changes of y. The consumed values condition is very strong; its consequences are discussed in section 2.9.4.

Consider again the AND gate shown in figure 2.5. After setting a to true, this value is not consumed before the next change of a. Hence, consumed values is violated, which means that the program is not guaranteed to be delay insensitive.

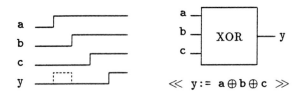

Figure 2.6: A three input XOR gate

Correspondence

The consumed values condition, by itself, is not sufficient to guarantee that a program is delay insensitive. In this section, we present two programs that satisfy consumed values but are not delay insensitive. Based on these, we state a second implementation condition for delay insensitive circuits, correspondence. In the next section, we will show that these two conditions are sufficient.

Consider the XOR gate shown in figure 2.6. We note that whenever the transition instance for an XOR gate has an active execution, it consumes the values of every input variable (a, b, and c in figure 2.6). This is because if any *single* input had the opposite value, the expression a $\oplus$ b $\oplus$ c would also have the opposite value.

Now consider the scenario depicted in figure 2.6. If the a input is changed and the b input is changed slightly later, the y output may or may not output a high pulse. The behavior depends on the delay of the XOR gate and the time between the changes of the a and b inputs. However, this is not a violation of the consumed values condition. When the c input eventually becomes true the transition instance becomes active again. At the subsequent execution, all the values of three inputs are consumed.

The problem with the above example is that the transition instance for the XOR gate can be disabled while it is active. This suggests the following guess for the second implementation condition:

> *Initial guess:* No variable read by a transition instance may be changed while the transition instance is active.

A similar condition has been proposed by Chandy and Misra [1]. This condition forbids examples like the one above with the XOR gate. When the a input became true, all inputs would have to remain stable until after y became true, i.e. the pulse would always be generated.

Our initial guess is not strong enough. Wire delays can also alter the order of arrivals of inputs. Consider a two input gate. If in the original program, the inputs go through the sequence $(F, T) \rightarrow (F, F) \rightarrow (T, F)$ the

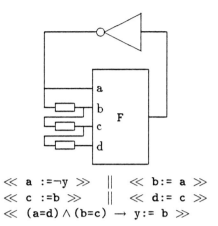

<< a := ¬y >> || << b := a >>
<< c := b >> || << d := c >>
<< (a=d) ∧ (b=c) → y := b >>

Figure 2.7: Example for the correspondence condition

combination (T, T) is never seen. However, if a wire delay is introduced to the second input, its change to false may arrive at the input of the gate after the first input becomes true. In this case, the input (T, T) may be observed; if this activates the transition instance for the gate, the program with the delay can function differently than the original program. The next example is a complete program demonstrating that this problem can occur in a program that satisfies the consumed values condition. Consider the program of figure 2.7. This program satisfies the following invariant:

$$
\begin{array}{l}
(\;a = \;\;b = \;\;c = \;\;d = y)\;\;\vee \\
(\neg a = \;\;b = \;\;c = \;\;d = y)\;\;\vee \\
(\neg a = \neg b = \;\;c = \;\;d = y)\;\;\vee \\
(\neg a = \neg b = \neg c = \;\;d = y)\;\;\vee \\
(\neg a = \neg b = \neg c = \neg d = y)
\end{array}
$$

which implies $(y = b \Rightarrow b = d)$. However, if the delay from the output of the inverter to the a input of the F block is sufficiently longer than the delays to b and c, then they may change before a does. This satisfies the precondition of the last transition. Executing the transition at this point brings the circuit to a state not satisfying the invariant. Thus, the program does not correspond to a delay insensitive circuit. However, the program does satisfy consumed values and the initial guess for an implementation condition given above. The problem with this program is that *delaying* the change of an input to a transition instance allows that instance to become active *sooner*.

We want to ensure that wire delays do not cause transition instances to become unintentionally active. Due to wire delays, a transition instance may see the values of the variables it reads change in an arbitrary order between two consecutive executions. Therefore, an implementation condition should guarantee that regardless of the order in which these changes occur, the transition instance never accidently becomes active. We will now formulate such a condition.

Let P be a program that satisfies the consumed values condition, let t be any transition instance in P, and let x be any variable read by t. Since P satisfies consumed values, x changes at most once between two times that t is active and executed. Therefore, the set of variables read by t that have a different value (compared to the last time t was executed) grows monotonically. To make sure that t never accidently becomes active, it is sufficient to record in an auxiliary variable the values of the variables read by t each time t is active and executed, and check that no combination of the old values and the current values activates t. This is now defined more rigorously.

For each transition instance t, let S^t be a vector valued auxiliary variable with one component for each variable read or written by t. $S^t_{\dim}$ denotes the dimensionality (number of components) of S^t. We write S^t_{now} to denote the value of this vector in the current state of the program and S^t_{last} to denote the value at the last completion of an active execution of t. $span^t(S^t_{\text{last}}, S^t_{\text{now}})$ is the set of all vectors that can be obtained by combining values from S^t_{last} and S^t_{now}. In particular,

$$p \in span^t(S^t_{\text{last}}, S^t_{\text{now}}) \; \textit{iff} \; \forall i \in \{1, \ldots, S^t_{\dim}\} : (p[i] = S^t_{\text{last}}[i]) \vee (p[i] = S^t_{\text{now}}[i])$$

Points at which t is active are called **activation points**. We write $active^t(p)$ to denote that p is an activation point of t. We can now define the correspondence condition.

Definition 4 *A program P satisfies* **correspondence** *if and only if for each transition instance, t and each $p \in S^t$*

$$p \in span^t(S^t_{\text{last}}, S^t_{\text{now}}) \wedge active^t(p) \Rightarrow p = S^t_{\text{now}}$$

Both examples in this section can be shown to violate correspondence.

2.9.2   Soundness of implementation conditions

The consumed values and correspondence implementation conditions were motivated by considering programs for circuits that *were not* delay insensitive. In this section, we sketch a proof that consumed values and correspondence are sufficient conditions: any program that satisfies these two is delay insensitive. A detailed presentation is given in [6]. The proof has two parts corresponding to properties $R.1$ and $R.2$ given in section 2.9:

1. We add a wire transition to the original program and show that the augmented program implements the original.

2. We show that the augmented program also satisfies the implementation conditions.

To show that the augmented program implements the original, we must show that for each execution of a transition instance of the augmented program either: (1) the variables of the original program were unchanged, or (2) there exists a corresponding execution of a transition instance in the original program. Let $t_{x'}$ be the transition instance in the augmented program that reads x', and t_x be the corresponding transition instance in the original program. Let t_w denote the wire transition ($\ll$ x' := x $\gg$).

Consider, t_c, an active transition instance executed in the augmented program. If $t_c = t_w$, it only modifies x'; therefore, none of the state variables of the original program are modified. Otherwise if $t_c \neq t_{x'}$, performing the corresponding transition instance in the original program will have the same effect. Only $t_{x'}$ remains. We will show that if $t_{x'}$ is active, then x' = x. For the sake of contradiction, assume otherwise. Then $t_{x'}$ is active, and x' $\neq$ x. This means that the value of x must have changed since the last time $t_{x'}$ was executed. Therefore, x' has the old value of x. It follows that $S_{now}^{t_{x'}}$ is a combination of values from $S_{last}^{t_x}$ and $S_{now}^{t_x}$. This would mean that the correspondence condition was violated in the original program, a contradiction. Therefore, x' = x whenever $t_{x'}$ is active, and executing t_x in the original program has the same effect.

We have now shown that every execution of the augmented program corresponds to some execution of the original program. Since the original program satisfies the implementation conditions, we only need to consider t_w to show that the augmented program also satisfies them. The only subtle argument is to show that t_w has consumed the old value of x before x is changed. Since the original program satisfies consumed values, t_x consumes the each value of x; this means that t_x is active and executed at least once between modifications of x. By the argument that the augmented program implements the original, we conclude that $t_{x'}$ is executed at least once between modifications of x, and when it is executed, x' = x. This implies that t_w is executed at least once between modifications of x, and the consumed values condition is satisfied.

In this section, we have argued that the consumed values and correspondence conditions are sufficient to guarantee delay insensitivity. In the remainder of this chapter, we will show how these implementation conditions can be employed to show that a program for a FIFO is delay insensitive. We will then present some of the consequences of these two implementation conditions and show that the class of programs that satisfy them is very small. We will show how this leads to a systematic approach to designing delay insensitive circuits.

2.9.3 A delay insensitive FIFO

Consider again a FIFO queue (see also section 2.8.5). A delay insensitive program for a FIFO is shown in figure 2.8. This program is on a rather abstract level. For example it uses a type **element** which includes the designated value E ("empty") along with the values for data in the FIFO. To realize this type in digital hardware, **element** must be implemented as an ensemble of bits. Empty values separate successive data values in the queue. In a stable configuration, the nodes at the "tail" of the queue all hold the same value, and the nodes at the "head" alternate between empty and non-empty elements as illustrated below:

$$q[1] \rightsquigarrow q[2] \rightsquigarrow \ldots q[i-2] \rightsquigarrow q[i-1] \rightsquigarrow Q[i] \rightsquigarrow q[i+1] \rightsquigarrow Q[i+2] \rightsquigarrow \ldots q[n-1] \rightsquigarrow Q[n]$$

Empty nodes are shown in small letters and full nodes in capitals.

Input of a non-empty value can occur only when q[1] = q[2] = E. Similarly, input of an empty value can occur only when the first two nodes hold a non-empty value. When a new datum is applied to the input of the queue (q[1] $\neq$ E), it is propagated to the right until it reaches a stage whose successor's output is not empty. In the above example, a new, non-empty input, would propagate to stage q[i-2]; it would not continue to q[i-1] because Q[i] is non-empty. Likewise, an empty input value will propagate to the right until it reaches a stage whose successor's output is empty. Successive inputs must alternate between empty (E) and non-empty data values.

Each queue stage alternates between holding an element ($\neq$ E) and having the empty value (E). This alternation makes it possible to detect changes that signal the completion of storing an element in a stage.

Proving that the FIFO is delay insensitive

It is now shown that the FIFO meets the two implementation conditions. Let t^i be the transition that writes q[i]:

$$\ll \ (\text{q[i-1]=E}) \neq (\text{q[i+1]=E}) \ \rightarrow \ \text{q[i]} := \text{q[i-1]} \ \gg$$

To show that the program satisfies the implementation conditions, two invariants I_1 and I_2 are useful. The invariants are:

$$I_1 : \ \forall i : \ \mathbf{a}^{\mathbf{q}[i]} = \{\text{q}[i] \neq \text{q}[i-1] \mid t^{i-1}\} \cup \{\text{q}[i] = \text{q}[i+1] \mid t^{i+1}\}$$
$$I_2 : \ \forall i : \ (\text{q}[i] = \text{q}[i+1]) \Leftrightarrow ((\text{q}[i] = E) = (\text{q}[i+1] = E))$$

Where the $\mathbf{a}^{\mathbf{q}}$ are auxiliary variables as described in section 2.9.1, and the notation $\{\text{q}[i] \neq \text{q}[i-1] \mid t^{i-1}\}$ denotes a set that contains t^{i-1} if q[i] $\neq$ q[i-1] and is empty otherwise (by analogy with section 2.2.5). Invariant I_1 relates the values of the auxiliary variables $\mathbf{a}^{\mathbf{q}}$ to the values of the state variables q. I_2 ensures that if two consecutive elements are non-empty ($\neq$E) then they are

```
CELL fifo2;
   STATE q: ARRAY[1..2*n] OF element;
   BEGIN
     {||i:[2..2*n-1] | ≪(q[i-1]=E) ≠ (q[i+1]=E)→q[i]:=q[i-1]≫}
   END fifo2;
```

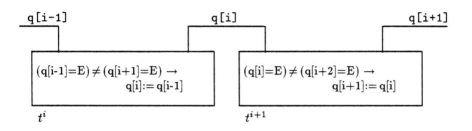

Figure 2.8: A delay insensitive FIFO

equal. These two invariants have been verified with LP using the techniques described in section 2.7 (see also [21]).

From the invariants, it is straightforward to show that the implementation conditions are met. When t^i is active, q[i] is written. To satisfy "consumed values," it must be shown that

$$\mathtt{a}^{\mathtt{q}[i]} = R^{\mathtt{q}[i]}$$

when t^i is active. For t^i to be active, $\mathtt{q}[i] \neq \mathtt{q}[i-1]$. From I_2 and the precondition of t^i we have $\mathtt{q}[i] = \mathtt{q}[i+1]$ before executing the transition instance. From I_1, we then conclude

$$\mathtt{a}^{\mathtt{q}[i]} = \{t^{i-1}, t^{i+1}\} = R^{\mathtt{q}[i]}$$

Thus, consumed values is satisfied.

To show that correspondence is satisfied, we first show that each of the q's alternates between empty and non-empty values. If t^i is active, then q[i].*pre* ≠ q[i − 1].*pre*. From I_2, we conclude (q[i].*pre* = E) ≠ (q[i − 1].*pre* = E). By the action of t^i, we have q[i].*post* = q[i − 1].*pre*, and therefore (q[i].*pre* = E) ≠ (q[i].*post* = E) as desired.

Since q[i] alternates between empty and non-empty values, we conclude, by the text of t^i, that the values of q[i-1] and q[i+1] must also change between empty and non-empty values between successive activations of t^i. We have already shown that consumed values is satisfied; therefore, q[i-1] and q[i+1] change exactly once between successive activations of t^i. We conclude that there are exactly four points in the span of S^{t^i} when t^i is active. The first point is the old completion point (which is not an activation point because q[i] = q[i-1]), two of these points correspond to states where only one of q[i-1] and q[i+1] have changed (and therefore t^i is not enabled), and the final point is the current activation point. Therefore, correspondence is satisfied.

The program shown in figure 2.8 meets both implementation conditions. Therefore, a realization of this program is delay insensitive with respect to the state variables named in the program. In particular, the FIFO will function correctly with arbitrary delays in the execution of the transition instances that update the q[i]'s and with arbitrary delays along the wires that distribute these values.

A Hardware realization

To realize the FIFO with a circuit, the element type must be encoded. This leads to a more detailed description of the circuit for each node. For simplicity, assume that element has values of T, F, and E. An element, e, can be represented with two boolean values, e.t and e.f (a **double-rail code**), as shown below:

e.t	e.f	e
FALSE	FALSE	E
FALSE	TRUE	F
TRUE	FALSE	T
TRUE	TRUE	illegal

For the one-bit FIFO, each queue element can be represented with two wires. The Synchronized Transitions description is shown below:

```
CELL fifo3;
  TYPE
    element = RECORD
        t,f: BOOLEAN;
    END;
  CELL t(in, out, succ: element);
  BEGIN
      ≪ ¬(in.t ∨ in.f) ≠ ¬(succ.t ∨ succ.f) → out:= in ≫
  END t;
  STATE q: ARRAY[1..2*n] OF element;
BEGIN
  { ‖ i: [2..2*n-1] | t(q[i-1], q[i], q[i+1]) }
END fifo3;
```

This program is a simple transformation of the program considered in the previous section, and by similar arguments, it can be shown to satisfy the implementation conditions.

To realize the program we need circuits to implement t and wires to connect these circuits together. In general, the circuits implementing a transition must be **monotonicity preserving**: if all inputs change monotonically, the output must also change monotonically[5]. If we are provided with a monotonicity preserving realization of t then the circuit corresponding to fifo3 is delay insensitive, and it will operate correctly regardless of delays on the wires corresponding to q.

We now consider an implementation of t as shown in figure 2.9. Each of the transitions in this program has a straightforward hardware realization. The transition,

$$\ll \text{ TRUE } \rightarrow \text{ temp:= } \neg(\text{succ.t } \vee \text{ succ.f}) \gg$$

corresponds to a NOR-gate, and the transitions

$$\ll \text{ in.f = temp } \rightarrow \text{ out.f:= in.f } \gg$$
$$\ll \text{ in.t = temp } \rightarrow \text{ out.t:= in.t } \gg$$

correspond to Muller C-elements (the output of a C-element is set to the last value at which the inputs are the same). The realization of t must be monotonicity preserving. To show this, further details of the realization are needed, e.g., the transistor level realization of the NOR and C gates. We will not go into these details here, but note that the reasoning needed to demonstrate that t is monotonicity preserving is completely local, it does not depend on details about other components or their interconnection.

Furthermore, it is important to note that t may not be internally delay insensitive. For example, only one of out.t and out.f changes when in is propagated. Accordingly, each value of temp is consumed by either the transition instance which writes out.t or the one which writes out.f, *but not both*. This illustrates an important point:

- Proving that a design is delay insensitive at one level of description *does not* prove that it is delay insensitive at lower levels.

This is not surprising, since the higher level description does not specify details of the lower level program.

At first, it may seem disappointing that the FIFO realization is not absolutely delay insensitive. The class of absolutely insensitive circuits is very

---

[5] The requirement of monotonicity provides a mapping between physical quantities (e.g. voltages) of the circuit and the logical values (e.g. boolean variables) of the program. For a more extensive discussion, see [24].

```
CELL t(in, out, succ: element);
  STATE temp:  BOOLEAN;
BEGIN
  ≪ TRUE  →  temp:= ¬ (succ.t ∨ succ.f) ≫      (* NOR gate *)
  ≪ in.f = temp  →  out.f:= in.f ≫            (* C-element *)
  ≪ in.t = temp  →  out.t:= in.t ≫            (* C-element *)
END t;
```

Figure 2.9: A FIFO Stage

small, as will be shown in the next section. Therefore, practical designs require cells that are internally delay sensitive. Fortunately, the critical delays can be contained in small modules (like the FIFO stage) where other verification approaches (typically more informal) are feasible. The consumed values and correspondence conditions specify how such cells can be composed such that there are no restriction on the delays of wires between such cells. This provides a flexible framework for designing VLSI systems that are delay insensitive down to the level of very small, simple cells.

Given a Synchronized Transitions program, there is a very clear distinction between what signals are assumed to be isochronic and which are delay insensitive. *Signals corresponding to state variables explicitly mentioned in the program are delay insensitive* if the program meets the two implementation conditions. Nothing can be said about signals not named in the program, e.g., internal wires used within the circuit realizing one transition instance.

2.9.4   Consequences of the implementation conditions

The consumed values and correspondence conditions are very strong. In this section, we present some consequences of these conditions. These consequence show that to implement any non-trivial function, it is necessary to use circuits that are not purely delay insensitive. Just as the FIFO stage presented in the previous section was internally delay sensitive, so are most practical circuits.

However, it is possible to build circuits where only very localized assumptions about delays are needed, just as in the FIFO stage. We will then show how the program for the FIFO provides a template for designing circuits that are delay insensitive down to the level of cells that only contain a few cells.

Circuit elements that are never used again after a finite number of operations are the exception rather than the rule in typical hardware designs. Accordingly, we assume that no transition instance has a last active execution, and that no signal has a last change. Furthermore, we are interested in *digital* systems; therefore, we assume that all variables are boolean valued. Other types (e.g. double-rail codes, integers, etc.) can be represented by ensembles of boolean variables. Throughout the following, P denotes a program that satisfies consumed values and correspondence, t denotes an arbitrary transition instance in P, and x and y denote an arbitrary state variables of P.

If t reads y, then t must have both an activation point where y = TRUE *and another where* y = FALSE. We assumed above that y continues to change. Since the set of transition instances R^y is the same for all values of y, each these transition instances, including t, must consume all values of y. This excludes the following transitions:

$$\ll \quad y \ \rightarrow \ \ldots \gg$$
$$\ll \ \neg y \ \rightarrow \ \ldots \gg$$
$$\ll \quad y \wedge (\ldots) \ \rightarrow \ \ldots \gg$$

The sequence of computations performed by each transition instance t depends only on the initial state of the program. Let E and E' be two executions starting from the same state. Consider t_1, the first transition instance executed in E. If this is not the first transition instance executed in E', we note that it was active when E' started (because the two executions started from the same state), and by the correspondence condition, it remained active until it was executed. Furthermore, no transition instance dependent with t_1 (see section 2.6.2) was executed in E' before t_1. Therefore, t_1 could be moved to the front of E' without changing the computation performed by any transition. This argument can be applied inductively to show that the sequence E' can be reordered to form the sequence E without changing the sequence of actions performed by any transition instance. A similar result is presented in [16].

The action of a transition may be a multiassignment. To check delay-insensitivity at a lower level, such transitions can be divided into several transitions with single assignments. The next property applies to programs without multiassignments.

If t reads x and writes y, then x changes exactly once between two changes of y. From the consumed values condition, we know that x changes at most once between two active executions. Consider what happens if y changes twice followed by a change of x. To satisfy correspondence, x must have been stable between the two changes of y; otherwise x could have changed while t was active. We conclude that the *second* change of y somehow caused the change of x. In particular, there must be some sequence of state variables, $(y, s_1, \ldots s_n, x)$ and a corresponding sequence of transition instances $(t_1, \ldots t_{n+1})$ such that t_1 became active depending on the value of y after the second change of y and modified the value of s_1. This in turn activated t_2, and so on, until t_{n+1} was activated to modify x. Since y changed twice in a period with no change of x, some transition instance in this sequence had an input change twice with no change of its output. This is a violation of the consumed values condition.

This last consequence shows that any circuit which is delay insensitive at the gate level can only consist of inverters and C-elements. In fact, any program that is delay insensitive without assuming atomic multiassignments must be a simple oscillator! A similar result was described by Martin [12].

Generalizing the FIFO

The building blocks presented below are simple variations of the FIFO cell; together they provide a powerful basis for building delay insensitive circuits.

The FIFO element can be extended with a combinational function, which computes a new value to be output, instead of just copying the input as it is done in the FIFO.

$$\ll \ (\texttt{in=}E) \neq (\texttt{succ=}E) \ \rightarrow \ \texttt{out:=} \ F(\texttt{in}) \ \gg$$

The function F can be arbitrarily complicated and time consuming; however, there are some restrictions:

- F must be E-preserving: $F(x) = E$ if and only if $x = E$.

- The realization of F must be monotonicity preserving.

A sequence can be distributed to two different destinations with a split element

$$\ll \ (\texttt{in=E}) \neq (\texttt{succ1=E}) \wedge (\texttt{in=E}) \neq (\texttt{succ2=E}) \rightarrow \texttt{out1,out2:=} F1(\texttt{in}), F2(\texttt{in}) \gg$$

Conversely, two sequences can be merged with a join element:

$$\ll \ (\texttt{in1=E}) \neq (\texttt{succ=E}) \wedge (\texttt{in2=E}) \neq (\texttt{succ=E}) \rightarrow \texttt{out:=} F(\texttt{in1,in2}) \gg$$

Similar building blocks have also been proposed by [19]. Furthermore, an iterative computation can be done with a structure formed by connecting the head and tail of a pipeline of at least three stages. Christian Nielsen [17] has applied this approach to the design of a matrix-vector multiplier. The design is implemented with double-rail code signals. Simple two input gates for these double-rail code signals were internally delay sensitive; above this level, the design is delay insensitive.

Using "fork" and "join," a diverse variety of pipelined networks can be implemented. These designs satisfy the consumed values and correspondence conditions; therefore, they are delay-insensitive, and free from timing hazards or metastable conditions.

2.9.5   Related work

Many of the ideas of asynchronous switching circuits were first developed in the 1950's. For example, monotonic circuits are presented in [24]. In this period, the primary concern was gate delays. With today's VLSI technology, wire delays are significant. We define a design to be delay insensitive if the explicitly added wire delays (the wire transitions) do not affect the computation. Similar approaches have been used by several others [14, 16, 23].

Our investigation of delay insensitive circuits was inspired by a self-timed division chip [25]. This chip demonstrated that delay insensitive circuits can provide a performance advantage over clocked implementations. Each iteration can begin as soon as the previous is finished; whereas, a clocked design must wait for the worst possible delays in every cycle.

In his recent Turing lecture [22], Sutherland presented "transition signaling" as an approach for building self-timed FIFOs and pipelines and gave numerous examples. His designs depend on matching the delays in data paths to delays in the corresponding control circuitry. However, the control circuitry is very similar to the designs presented in this paper.

At Caltech, Alain Martin and his colleagues have explored another approach to designing delay insensitive circuits which is also based on parallel programming. They have obtained impressive results including the design of a simple self-timed microprocessor[13]. Their approach is based on a specification in a high-level language using message passing and applying *systematic transformations* to obtain the final design. Our emphasis has been a complementary one which stresses issues of *verification* and formal models of delay insensitive circuits.

In [1], Chandy and Misra described a class of unity programs thay they called "delay insensitive." These programs correspond the class identified as speed-independent programs at the beginning of this section. Chandy and Misra proposed a condition for this class. Their conditions requires that the value of expressions on the right hand side of assignments do not change while the transition is active. Since delay insensitive circuits are a subset of speed-independent circuits, our implementation conditions should imply Chandy and Misra's condition. Indeed, their condition is implied by the correspondence condition.

2.10   Conclusion

In this chapter, we have described a high level design language for developing VLSI circuits. The same source program is used for both verification and synthesis. Here, the verification aspects have been emphasized. In particular, the importance of mechanical verification has been stressed.

It is possible to design both synchronous and asynchronous circuits using our approach. The same high level language and verification techniques are used for both. We have given a formal definition of delay insensitivity and a way of building such circuits.

Acknowledgement

The development of Synchronized Transitions has received help from many sources. Financial support has been given by Århus University, The Danish Technical Research Council and The Danish Natural Science Research Council. We would like to thank the Computer Science Department at Århus University where much of the early work on Synchronized Transitions was done. We are also grateful to Kai Li and Ken Steiglitz for their support and encouragement of this research at Princeton.

Anders P. Ravn provided valuable insight and inspiration in the development of the Synchronized Transitions notation.

The work on automatic verification has been done in cooperation with John Guttag and Steve Garland at MIT.

The use of abstraction functions and the abstraction condition was inspired by Anders Gammelgaard's thesis from Århus University.

Susan Owicki suggested the reordering method to show that programs satisfying the consumed values and correspondence conditions are deterministic.

Our approach to the design of delay insensitive circuits has been strongly influenced by Ted Williams and the VLSI club at Århus University.

Niels Mellergaard wrote the translator which forms the basis of all our tools.

Finally, we would like to thank Anders P. Ravn, Anders Gammelgaard, Hartmut Schmeck, Steven Garland, John Guttag, Christian D. Nielsen and Henrik Hulgård for their very careful reading of preliminary versions of our papers.

2.11   References

[1] K.M. Chandy and J. Misra. *A Foundation of Parallel Program Design.* Prentice Hall, 1987.

[2] European Silicon Structures, Bracknell, Berkshire RG12 3DY, United Kingdom. *Designing in MODEL*, doc. no. ES2-014-0055 edition. SOLO Reference Manual.

[3] Stephen J. Garland and John V. Guttag. Inductive methods for reasoning about abstract data types. In *Proceedings of the 15th ACM Conf. on Principles of Programming Languages*, 1988.

[4] Stephen J. Garland and John V. Guttag. An overview of LP: the Larch Prover. In *Proceedings of the Third International Conference on Rewriting Techniques and Applications*. Springer-Verlag, 1989.

[5] Stephen J. Garland, John V. Guttag, and Jørgen Staunstrup. Verification of VLSI circuits using LP. In *The Fusion of Hardware Design and Verification*. IFIP WG. 10.2, North Holland, 1988.

[6] Mark R. Greenstreet and Jørgen Staunstrup. Atomicity, programs, and hardware. Technical Report ID-TR: 1988-46, Department of Computer Science, Technical University of Denmark, Lyngby, Denmark, November 1988.

[7] John V. Guttag and Jim J. Horning. Report on the Larch Shared Language and a Larch Shared Language handbook. *Science of Computer Programming*, 6(2):103–157, March 1986.

[8] D.D. Hill and D.R. Coelho. *Multi-level Simulation for VLSI Design*. Kluwer Academic Publishers, 1986.

[9] C.A.R. Hoare. *Communicating Sequential Processes*. Prentice-Hall International Series in Computer Science, 1985.

[10] D.E. Knuth and P.B. Bendix. Simple word problems in universal algebras. In J. Leech, editor, *Computational Problems in Abstract Algebra*, pages 263–297. Pergamon Press, 1969.

[11] Leslie Lamport and Fred B. Schneider. The 'Hoare logic' of CSP, and all that. *ACM Transactions on Programming Languages*, 6(2), April 1984.

[12] Alain J. Martin. The limitations to delay-insensitivity in asynchronous circuits. Technical Report Caltech-CS-TR-90-02, Computer Science Department, California Institute of Technology, Pasadena, CA 91125, January 1990.

[13] Alain J. Martin et al. The design of an asynchronous microprocessor. In *Proceedings of the Conference on Advanced Research in VLSI*, Caltech, 1989.

[14] C.E. Molnar, T.P. Fang, and F.U. Rosenberger. Synthesis of delay-insensitive modules. In H. Fuchs, editor, *Proceedings of the 1985 Chapel Hill Conference on VLSI.* Computer Science Press, 1985.

[15] J.D. Morison et al. ELLA: Hardware description or specification. In *Proceedings of 1984 IEEE ICCAD*, 1984.

[16] D.E. Muller and W.S. Bartkey. A theory of asynchronous circuits. In *Proceedings from International Symposium on the Theory of Switching.* Harvard University Press, 1959.

[17] C.D. Nielsen. Design of delay insensitive circuits using synchronized transitions. Master's thesis, Department of Computer Science, Technical University of Denmark, 1990.

[18] Susan Owicki and David Gries. An axiomatic proof technique for parallel programs I. *Acta Informatica*, 6(1), 1976.

[19] N.P. Singh. A design methodology for self-timed systems. Master's thesis, Laboratory for Computer Science, MIT, 1981. Technical Report TR-258.

[20] Jørgen Staunstrup, S. Garland, and J. Guttag. Localized verification of circuit descriptions. In *Proceedings of the Workshop on Automatic Verification Methods for Finite State Systems, LNCS 407.* Springer Verlag, 1989.

[21] Jørgen Staunstrup and Mark Greenstreet. Designing delay insensitive circuits using 'Synchronized Transitions'. In Luc J.M. Claesen, editor, *Formal VLSI Specification and Synthesis. VLSI Design Methods*, volume 1, pages 209–226. North-Holland/Elsevier, 1990.

[22] I.E. Sutherland. Micropipelines. *Communications of the ACM*, 32(6), June 1989.

[23] J.T. Udding. *Classification and Composition of Delay-Insensitive Circuits.* PhD thesis, Eindhoven University of Technology, 1984.

[24] Stephen H. Unger. *Asynchronous Sequential Switching Circuits.* Wiley-Interscience, 1969.

[25] T.E. Williams, M. Horowitz, et al. A self-timed chip for division. In *Proceedings of the Conference on Advanced Research in VLSI.* Computer Science Press, 1987.

Formal Methods for VLSI Design
J. Staunstrup (Editor)
Elsevier Science Publishers B.V. (North-Holland)
© IFIP, 1990

Chapter 3

Verifying SECD in HOL

G. Birtwistle and B. Graham[1]

Abstract

This paper describes some of the work done at Calgary on the design of an SECD chip and its verification using the Cambridge HOL proof assistant. The chip is a physical realization of Henderson's variant of Landin's abstract architecture to execute the lambda calculus. The machine uses closures and includes explicit machine instructions to assist recursion. The complete proof, which goes from an abstract specification down to the transistor level, is far too involved to be covered in a single paper. In this paper, we discuss the SECD architecture and design and trace through a portion of the proof of correctness of one sequence of the microcode.

3.1 Introduction

Work on SECD started in 1985. The aim of the project was to investigate the specification driven design methodology (design for verifiability) by designing a reasonably substantial chip and verifying it with the HOL proof assistant [10]. We chose SECD because it was well known and well understood and there was much work to build upon — see for example, [16] and [22]. We

---

[1] Addr.: Department of Computer Science, University of Calgary, 2500 University Drive, Calgary, Alberta, Canada T2N 1N4

chose HOL because it was there and had the best track record in hardware verification. Besides the proof, we have also: (i) designed, fabricated and are testing a version the chip, (ii) constructed a rig and associated software (including a compiler) so that we can compile Lispkit[2] programs and download them to run on SECD, (iii) completed a (hand) proof that the abstract SECD machine executes Lispkit programs correctly.

The SECD chip arose within an ongoing research effort by the VLSI and verification group at the University of Calgary. The chip was used as a vehicle to explore the use of specification to drive design synthesis. The methodology entails elaborating a design hierarchically as a tree of nodes and formally specifying the behaviour at each node. Verifying that the composition of behaviours of a node's children agrees with the node's specification assures a correct design. By deductive argument, we can show the correctness or otherwise of a complete design. Thus the design and verification process is the object of study, and the chip is only a byproduct providing the team with hands-on experience with a nontrivial design. This intention gave rise to design criteria (make it simple, testable, useful, but above all correct) that affected decisions taken throughout the design.

The perceived need for higher levels of reliability of computer and control systems has driven the exploration of formal verification methods. Software and hardware verification efforts have generally been undertaken independently. A common notation for all levels of hardware design as well as the software running on the systems is advantageous when attempting to verify a complete system. The HOL notation used in our work is sufficiently expressive to achieve this objective.

The difficulty of verifying program correctness for imperative languages has motivated the choice of functional programming languages for our subject system. The strong mathematical basis of functional languages makes them more amenable to formal correctness study than imperative languages. Thus the choice of an architecture that supports functional programming was made with the intention that a completely verifiable system could be produced. SECD was chosen as a well understood and well documented architecture, and we had available a compiler for LispKit down to SECD code. We began with the assumption that the compiler can be shown to be correct, and that SECD executing compiled LispKit programs correctly implements the high level language.

The complete proof of SECD in HOL is far too large to be presented in a single paper. In other accounts, we have given the correctness proof of the translation schema from LispKit down to SECD [24], described the

---

[2] Lispkit is a pure functional Lisp language subset, see Henderson's text [16].

rig for running LispKit programs on SECD [26], documented the LispKit
to SECD compiler [23], described the abstraction mechanisms between the
various design levels of SECD [11] and explained the chip design [2], [12],
[13], [14]. In the rest of this paper we sketch the HOL notation and give an
introduction to carrying out proofs with HOL, describe the abstract SECD
machine and explain its formal specification, sketch the design of the SECD
chip and carry through a detailed look at a section of the proof of correctness
of the SECD microcode. In a companion paper in this volume, Lars Rossen
gives an account of another HOL based project — expressing Mary Sheeran's
Ruby in HOL.

3.2   Introducing HOL

The verification work in the SECD project was carried out in the HOL system
designed and written by Mike Gordon at the University of Cambridge. HOL
implements a version of Church's typed higher order logic [3]. The main bonus
of working with a higher order logic is that they permit clear, readable, and
succinct specifications. Higher order features are particularly useful when it
comes to joining sub-systems together and for describing architectures and
abstractions. The downside is that with increased expressive power, less can
be automated. In fact the HOL system is a proof-checker or proof-assistant
and not an automated theorem prover. You have to supply to HOL the
detailed working of what you think is a proof of the theorem at hand and
HOL checks things out step by step. Well documented examples of HOL can
be found in [4], [5], [6], [7], [8], [22] and [25].

The HOL logic is a conventional higher order logic. The HOL logic is
interfaced to the ML language in which proof strategies can be encoded. ML
has a very strong type discipline which is used to ensure that the only way
to create a theorem is to prove it: they cannot be faked. Term and theorem
are types in the implementation. The basic axioms are theorems and the only
means of producing new theorems is through inference rules. Theorems are
prefixed with a turnstile ⊢. The set of terms in basic HOL is defined by:

1. the bool constants T (true) and F (false) are terms

2. variables are terms

3. if R and S are terms, then so are $\sim R$ (negation), $R \land S$ (conjunction),
   $R \lor S$ (disjunction), $R ==> S$ (implication), $R <=> S$ (equivalence),
   and (R).

4. if x is a variable and *body* is a term, then *! x . body* (! stands for ∀),
   ? x . body (? stands for ∃), *@ x . body* (@ stands for the select operator)[3],
   and *\ x . body* (\ stands for λ) are terms.

5. a function application of the form *R S* where *R* and *S* are terms (the
   rator and rand respectively) is a term

6. a conditional *b => R mid S* where *b*, *R* and *S* are terms is a term.
   Read this as *if b then R else S*.

7. a local declaration *let x = expr in body* where *x* is a variable and *expr*
   and *body* are terms is a term. The local declaration is taken as a sugared
   version of *(\ x . body) expr*

The HOL system has a theory (library) facility where definitions and theorems can be stored. The HOL system comes equipped with several built-in
theories. These include theories for numbers (0, 1, 2, 3, ...), lists, and pairs.
The theories include the expected operators, e.g. $+$, $-$, $*$, ..., $=$[4], $<$, ...,
CONS, NIL, HD, TL, FST, SND. HOL comments are enclosed in % signs.

All terms in HOL are typed and the type of a term is usually inferred by
the system. The type of a constant is manifest. The type of a variable is
defined by its usage, e.g. in *a* ∧ *b = 5*, *a* is of type *bool* and *b* is of type
num. The type of a function from * to ** is written * → **, the type of a
list whose members are all of type * is * *list*, and the type of a pair is written
* # **, where * is the type of its first component and ** the type of its
second component.

3.2.1   Conducting proofs in HOL

The HOL system is constructed from 5 axioms and 8 inference rules. New
theorems may be proved from the axioms and previous theorems by applying
inference rules. This method of proof, forward proof, suffers from two hand-
icaps: (i) it is usually quite hard to know where to start, and (ii) there is a
lot of bookkeeping associated with large proofs. HOL supports forward proof,
but also has facilities for backward proof. In a backward proof, we start by
supplying what we want to prove as a *goal*. Goals may be rewritten or split
into simpler parts by *tactics*. Tactics may be thought of as inverse inference
rules. A tactic splits a goal into subgoals but we are assured that if we can
prove each of the subgoals then there is a companion inference rule to that

---

[3] The select operator is a higher order version of Hilbert's ε-operator.
[4] The operator '=' is polymorphic

tactic which will prove the parent goal from these proofs. Suppose tactic T takes goal G to subgoals $g1$ and $g2$. Then there will be an inference rule R which takes theorems $\vdash g1$ and $\vdash g2$ to the theorem $\vdash G$. For example, there is an inference rule $IMP\_ANTISYM\_RULE$ which constructs the theorem $\vdash a = b$ from the theorems $\vdash a ==> b$ and $\vdash b ==> a$. Corresponding to $IMP\_ANTISYM\_RULE$ there is a tactic EQ_TAC which when applied to the goal $a = b$ produces two subgoals $a ==> b$ and $b ==> a$. The idea behind a backward proof in HOL is that we recursively apply tactics to decompose a goal into a tree of subgoals until all the leaves are simple enough to prove. Then the HOL system will take our manually generated pattern of tactics and subgoals and transform it into a forward proof. HOL supplies several functions to manipulate goals. $g = set\_goal$ is used to initialise a goal stack to the next theorem to be proved; $e = expand$ takes a tactic and applies it to the current goal; $b = backup$ backs up the stack (throws away the last application of a tactic) and is very useful during proof explorations. If a tactic splits a goal into more than one goal (as with EQ_TAC) the subgoals are tackled one at a time. When one subgoal has been proved, the next to be proved is automatically popped.

3.2.2 Prove $\vdash (\exists\, x\,.\,(x = t) \wedge A\, x) = A\, t$

As an example, here is the proof in HOL of a primitive version of a theorem which is used to remove hidden wires in hardware proofs. We start by setting the goal and then split it into two implications using EQ_TAC.

```
#g "(? x:bool . (x = t) /\ A x) = A t";;
"(?x. (x = t) /\ A x) = A t"

#e(EQ_TAC);;
OK..
2 subgoals
"A t ==> (?x. (x = t) /\ A x)"

"(?x. (x = t) /\ A x) ==> A t"
```

It is easier to work with the implicands on the assumption list. We want this to happen for both subgoals so we back up and apply the compound tactic $EQ\_TAC\ THEN\ STRIP\_TAC$ to the original goal.

```
#b();;
"(?x. (x = t) /\ A x) = A t"

#e(EQ_TAC THEN STRIP_TAC);;
OK..
2 subgoals
"?x. (x = t) /\ A x"
    [ "A t" ]

"A t"
    [ "x = t" ]
    [ "A x" ]
```

The pertinent assumptions are listed under each subgoal. We work on the subgoals one at a time. The textually lower one takes precedence. We can use the two assumptions to attack the goal. A simple-minded method is to discharge $A \; x$ and then rewrite what we have with the assumption $x = t$.

```
#e(UNDISCH_TAC "(A:bool->bool) x");;
OK..
"A x ==> A t"
    [ "x = t" ]


#e(ASM_REWRITE_TAC []);;
OK..
goal proved
.  |- A x ==> A t
.. |- A t

Previous subproof:
"?x. (x = t) /\ A x"
    [ "A t" ]
```

ASM_REWRITE_TAC [ ] uses terms in the assumption list to perform rewrites, together with a few rules that are applied automatically. Extra theorems may be included explicitly in its argument list. The other subgoal is now popped. In this case we choose a suitable value for x (if we choose an inappropriate value, the goal will be rendered unprovable, but the system remains sound) and the proof is completed with a simple rewrite from the assumption list.

```
#e(EXISTS_TAC "t:bool");;
OK..
"(t = t) /\ A t"
    [ "A t" ]

#e(ASM_REWRITE_TAC []);;
OK..
goal proved
. |- (t = t) /\ A t
. |- ?x. (x = t) /\ A x
|- (?x. (x = t) /\ A x) = A t

Previous subproof:
goal proved
```

The session is completed by storing away the theorem in the current theory under the name *EXISTS_ELIM*. *prove_thm* is a built-in theorem which takes a save name (in single back quotes), a goal (in double quotes), and a tactic from which it generates a forward proof of the theorem.

```
#let EXISTS_ELIM = prove_thm
  ('EXISTS_ELIM',
   " ( ? x . (x = t) /\ ((A:bool->bool) x)) = A t",
   EQ_TAC THEN STRIP_TAC
   THENL
     [ UNDISCH_TAC "(A:bool->bool) x"
     ;
       EXISTS_TAC "t:bool"
     ]
   THEN ASM_REWRITE_TAC []
);;
```

Note that when a tactic splits a goal into several subgoals we supply suitable tactic for each of the subgoals in a list following a THENL. When each subgoal tactic has a common tail (here *ASM_REWRITE_TAC [ ]*) we can merge these subgoal branches.

3.2.3 Specifying hardware in HOL

Specification. A specification treats a device as a black box. Typically, it states what can be observed on each output line in terms of current and previous inputs. We express the signal value on line p at time t by $p\ t$, i.e. we take p as a function from $num \rightarrow bool$. Using this notation, here is the specification of a combinational exclusive or gate

```
! a b out . xor2 a b out = (! t . out t = ˜(a t <=> b t))
```

which is read as "the signals on lines a, b, and *out* are constrained by an xor2 gate to be in the relation that *out* at time t is high iff the signal values on a and b at time t differ".

The simplest sequential circuit is probably the unit delay which which simply delays the incoming signal by one time unit. The specification

```
! i out . del i out = (! t . out(t+1) = i t)
```

states that for all values of t ($t = 0, 1, 2, 3, ...$), *out(t+1)*, the signal value on port *out* at time $t+1$ equals that on the input line i at time t. The initial output value of this device is not defined. A variant on the theme is a clocked delay where the input is read only when its clock line goes high, otherwise the output remains the same (state is introduced). The specification of this device is

```
! i ck out . del2 i ck out = (! t . out(t+1) = (ck t => i t | out t))
```

We now specify the behaviour of a synchronous clocked register. ¿From clock tick to clock tick, the n–bit register may reset to zero, parallel load, or remain as it is. The register has an explicit clock line and remains unchanged unless this clock line goes high. The 3 control signals are handled by a pair of control wires *cntl*. We use the pair notation (c_0, c_1) to denote the value on the control line, and assign the following codes: $RESET = (T, T)$, $LOAD = (T, F)$, $NOP = (F, \_)$. We adopt the notation *out t k* to represent the value on the wire out_k at time t. It makes sense to bundle the input wires and the output wires together as buses. If we define a function $low\ k = F$ we can use the principal of extensionality to remove the indices k from the specification which then reads:

```
! n ck cntl i out . nReg n ck cntl i out
   = ! t . out (t+1) = ck t /\ cntl t = RESET => low
                     | ck t /\ cntl t = LOAD  => i t
                     | out t
```

Implementation. Implementations are much easier to define: we list the components and how they are wired together. As an example here is the HOL definition of the implementation of an xor2 gate constructed from 4 nand2 gates.

```
! a b out . xor2_imp a b out
   = ? r q p . nand2 a b p /\ nand2 a p q
             /\ nand2 p b r /\ nand2 q r out
```

The only difficulty comes in handling the internal wires. What we need is a notation that ensures that hidden lines are defined (and can be typed) and yet are not visible outside the definition. We also need a logical inference rule which allows us to remove all occurrences of hidden lines when we unfold implementation definitions. This is where stronger versions of the rule *EXISTS_ELIM* are used. When unfolding an implementation, we first rewrite with the definitions of the parts (nand2 in the example above). This puts the implementation definition in standard form, conjuncts of the form *output wire = expression* with all hidden lines existentially quantified. These hidden lines can then be (pruned) eliminated by applications of *EXISTS_ELIM*. Here is the way the definition of the xor2 implementation is unfolded and pruned:

```
xor2_imp a b out

= ? r q p . nand2 a b p /\ nand2 a p q
          /\ nand2 p b r /\ nand2 q r out

= ? r q p . p t = ~(a t /\ b t)  /\  q t  = ~(a t /\ p t)
          /\ r t = ~(p t /\ b t)  /\  out t = ~(q t /\ r t)

= ? r q . q t = ~(a t /\ (~(a t /\ b t)))
        /\ r t = ~((~(a t /\ b t)) /\ b t)
      /\ out t = ~(q t /\ r t)

=   out t = ~((~(a t /\ (~(a t /\ b t))))
            /\ (~((~(a t /\ b t)) /\ b t)))
```

We are left with an expression solely in terms of input and output values and ripe for simplification by the normal rules of logic.

To be a little more precise with HOL terminology, *UNFOLD*ing substitutes the expression defining the value of a signal for that signal throughout the rest of a term. *PRUNE*ing removes the quantified variable when its only remaining occurrence in the term is the left hand side of an equation.

3.3   The abstract SECD machine

Henderson's SECD machine has three load functions (one each for variables, constants, and in-line functions), *AP* to enter a non-recursive function call, *DUM* and *RAP* to enter recursive function calls, *RTN* for function exit, *SEL* and *JOIN* to support conditional branching, 4 built-in list operators, 7 comparison or arithmetic operations, and a halt instruction. We characterize the behaviour of an operating abstract SECD machine by giving its state at the end of an instruction. The state is a 4-tuple listing the contents of the 4 stacks that give it its name, *(S, E, C, D)*.

LD (m.n):	s e •.c d	⇒	(x.s) e c d
			where x = locate((m.n),e)
LDC x:	s e •.c d	⇒	(x.s) e c d
LDF c':	s e •.c d	⇒	((c'.e).s) e c d
AP:	((c'.e') v.s) e •.c d	⇒	nil (v.e') c' (s e c.d)
DUM:	s e •.c d	⇒	s (nil.e) c d
RAP:	((c'.e') v.s) (nil.e) •.c d	⇒	nil e" c' (s e c.d)
			where e" = replcar(v, e')
RTN:	(x) e' (•) (s e c.d)	⇒	(x.s) e c d
SEL cT cF:	(T.s) e •.c d	⇒	s e cT (c.d)
	(F.s) e •.c d	⇒	s e cF (c.d)
JOIN:	s e (•) (c.d)	⇒	s e c d
CAR:	((a.b).s) e •.c d	⇒	(a.s) e c d
ADD:	(a b.s) e •.c d	⇒	((b + a).s) e c d
STOP:	s e (•) d	⇒	s e (•) d
			Stop. Result is s.

CAR is typical of the unary operations; ADD is typical of the binary operations

Table 3.1: SECD code transitions

Broadly speaking, expressions are evaluated on *S*, *E* is the environment (a list of lists — each dynamically nested function introduces a new level of nomenclature, and several items may be defined at each level), and the SECD code is held in *C*. The old machine state *(S, E, C)* is pushed onto *D* when a new function is entered, and restored therefrom on return. The operation of the abstract SECD can therefore be specified by a transition table which shows how each instruction affects the state 4–tuple. For example under *AP* in table 3.1 we find *((c'.e') v.s) AP.c d ⇒ nil (v.e') c' (s e c.d)*. The compiler sees to it that prior to executing an *AP* instruction, we have placed the argument list *v* on the stack *s*, and also a function closure (the function code *c'* and its environment *e'* as a dotted pair on *s* above *v*). The effect of executing *AP* is to: set the stack *S* to empty; set a new environment by augmenting the defining environment for the function with the arguments to this call *(v.e')*; set the code pointer to *c'*, the function body; and to save *s e c* on the dump.

3.4   Designing the SECD

In this section we discuss transforming the abstract machine description into a microelectronic device. First we describe the chip interface, next the interpreter which was derived from the specification by pattern matching; then the register transfer level which was derived from the interpreter after basic architectural questions had been resolved. In the final subsection, we state the purposes of the 4 major floor plan elements in the SECD design.

3.4.1   The chip interface

Realising the SECD machine as a working system required that it be able to accept a task, compute a result, and return it. The interface to permit a user to pose a problem and the machine to return a result was the first major design decision. Two major options were considered: a co-processor role and a stand-alone system[5].

Using the SECD as a co-processor to another system would permit i/o to be handled by the other system rather than the SECD chip. For instance, using SECD as a co-processor for a SUN workstation would see the SUN able to read in an S-expression, set up a memory image for the problem, signal the SECD to begin computation, receive a signal back on completion of the calculation, and print out the S-expression solution. This has the advantage of simplifying the tasks the SECD must perform.

---

[5] This discussion summarizes work by Jeff Joyce on system configuration, reported originally in [17].

The second option considered was that of a stand-alone system. This would require incorporating primitive read and write operations into the definition of the machine. Ideally, an operating system for such a system would be written in the higher level LispKit language, but its pure functional nature provides recursion as the sole means to achieve the infinite 'while' loop required, but since every recursion uses up system resources, this would exhaust memory. A possible solution would be to hand modify the machine code to create the desired loop.

An extension of the stand-alone system concept is the multi-processor SECD system. Such a system was envisioned as having multiple (perhaps 100) SECD chips operating concurrently on shared memory. An operating system kernel would assign S-expressions to processors for evaluation. Each processor would be assigned exclusive use of memory blocks in a sort of virtual memory system, with blocks being assigned and garbage collected by the kernel.

The co-processor option was chosen as most appropriate to the scope of the project. If desired, a stand-alone approach could be used for a second iteration of the design.

Representation of S-expressions (the "stuff" of programs and data in the SECD machine) was not determined at this level, but it was known that they would be stored within a finite memory, and this necessitated garbage collection. In practice, a simple *mark and sweep* garbage collector was used. Thus, the SECD system consisted of a microprocessor operating upon an S-expression image in a RAM. Major phases in the operation of the system are:

- load problem into RAM (done by main processor)
- send *start* signal to SECD system
- run
- stop and signal completion to main processor
- return result (again, main processor task)

A four state finite state machine model was used to express the desired operating interface. Several ideas were incorporated including the addition of a *reset* input to permit a deterministic startup of the machine, an *idle* state for waiting until a problem is loaded, a *top_of_cycle* state as the top of a fetch/execute cycle, two *error* states, and a *button* input to control transitions from *error* and *idle* states. This view is summarized by Figure 3.1.

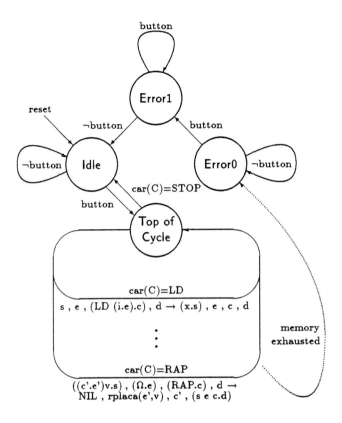

Figure 3.1: Top Level Finite State Machine View of SECD

3.4.2 The SECD interpreter

We designed SECD with ease of verification in mind. Since the total verification effort was known to be large, we settled on a simple architecture and a simple clocking strategy. We still underestimated the total verification effort by a large margin.

Perhaps the most interesting feature of the design effort was realising how much progress could be made by transforming the initial specification of SECD once a few architectural decisions had been made. Henderson [16] pointed out that it is possible to derive an interpreter for the SECD machine by noting how individual instructions update the *(S, E, C, D)* state 4-tuple as they are executed. We show how this is done for the *LDF* instruction, an instruction of medium complexity whose verification we will be tracking in the sequel.

$$\text{s e (\textbf{LDF} f).c d} \Rightarrow \text{((f.e).s) e c d}$$

A new computation is entered when an **LDF** instruction is carried out, The new computation builds a function closure *(c'.e)* on top of S finding the body of the function in-line, pairing it with the current environment *e*, and pushing the new pair onto S. The instruction *LDF f* is then peeled off the code sequence. E and D are not affected by the instruction. By pattern matching, *LDF = car(car C), f = car(cdr C)*, and the next instruction is located at *cdr(cdr C)*.

$$
\begin{aligned}
S &:= \text{cons(cons(car(cdr C), E), S);} \\
E &:= \text{E;} \\
C &:= \text{cdr(cdr C);} \\
D &:= \text{D;}
\end{aligned}
$$

In general we must take some care about the order in which we update the fields of the state 4-tuple. Here it is important to evaluate S before we evaluate C.

3.4.3   The register level view

We now refine this top level view into an architecture. For ease of implementation and verification, we settled on a single bus architecture and an external RAM memory M. The SECD stacks are registers pointing to addresses in the memory. The memory is initially chained together and referenced by another register *FREE*. Access to the external memory M is via a register *MAR*. A *read* operation places *M[MAR]* on the bus; a *write* operation stores the value on the bus in *M[MAR]*. We decided to go for 32-bit words in this implementation. With type tags taking 2 bits and garbage collection flags another 2 bits per word, that leaves space for either a 28-bit two's complement number or a cons'ed word containing two 14-bit addresses. This is a sufficient address space for running the LispKit compiler on the SECD machine itself.

We now refine the interpreter level instructions down towards the register-register level. Since *car*, *cdr*, and *cons* operations are very common we anticipate special hardware for them. At this stage we mechanically rewrite the operations for *LDF* into smaller steps, each step being of the form *R := rhs;* where R is a register and *rhs* is either a register, or a single *car*, *cdr* or *cons* operation on a register. Much of the time we are assembling sub-expressions in order to cons them. We simplify matters by allotting two general purpose temporary registers called *X1* and *X2*. When we are carrying out a *cons* operation, we put the second argument in *X2* and the first argument in *X1* (since we have a functional language, their order of evaluation is immaterial).

$$
\begin{aligned}
\textbf{S} &:= \textbf{cons(cons(car(cdr C), E), S)} \\
\text{X2} &:= \text{E;} \\
\text{X1} &:= \text{cdr S;} \\
\text{X1} &:= \text{car X1;} \\
\text{X1} &:= \text{cons X1 X2;} \\
\text{X2} &:= \text{S;} \\
\text{S} &:= \text{cons X1 X2;} \\
\textbf{C} &:= \textbf{cdr(cdr C)} \\
\text{C} &:= \text{cdr C;} \\
\text{C} &:= \text{cdr C;}
\end{aligned}
$$

In the main, each register-register operation is split into a read onto the bus and a write from the bus at the bus level. Each register X has two control signals associated with it, systematically called rX and wX. When rX goes high, the contents of X are read onto the bus. The controller sees to it that only one register may be read onto the bus at a time and for simplicity, only one register may be written from the bus at a time. When wX goes high, the contents of the bus are written into X: all the bits of the bus if X is a 32-bit register; if X is a 14-bit register, the top 18-bits of the bus are ignored except for the special CAR register which is wired to accept bits 14-27 of the bus. Bread and butter microcode can be generated directly from the register-register level description as soon as we supply a template for each class of assignment in the register-register level description.

- Copying the contents of register X to register Y is quite simply

$$\text{r}X \qquad \text{w}Y$$

Movement involving *cdr*, *car* or *cons* require access to the external memory. The first two cases are have simple and short templates; *cons* is rather more involved and is implemented as a micro-code subroutine.

- To carry out the request $Y := cdr\ X$ we place the contents of X in MAR, read $M[MAR]$ onto the bus (all 32 bits), and then store its *cdr* field in Y.

$$
\begin{aligned}
&\text{r}X &&\text{wmar} \\
&\text{rmem} &&\text{w}Y
\end{aligned}
$$

- To carry out the request $Y := car\ X$ we place the contents of X in
  MAR, read $M[MAR]$ onto the bus (all 32 bits), then filter out the *car*
  bits via the CAR register before passing the result to Y.

rX	wmar
rmem	wCAR
rCAR	wY

- The much used *cons* operation requires more thought. A call on *cons* requires the freeing of a fresh cell in memory (perhaps initiating a garbage
  collection) and the passing of two addresses to memory to fill out its
  fields. We have chosen to restrict *cons* to operating upon two specific
  registers which we call *X1* and *X2*. Our *consX1X2* operation locates
  a free cell in memory, fills out its two address fields from *X1* and *X2*,
  sets its type bits to indicate a cons'ed word, zeroes its garbage collection bits, writes it to the memory, and returns its address. Calls to
  consX1X2 occur so frequently in the code and are of such complexity
  that we inserted a sub-routine facility into the micro-code.

  if (null rfree) then GCBegin()

rfree	wmar
rmem	wfree
rcons	wmem

At the end of this analysis we have seen the need for two auxiliary registers
and micro-code routine for consing their contents and placing the address of
the new cell on the bus. Figure 3.2 summarizes the register-register level
view of the chip design. The control part of the design is shown as a classic
finite state machine, with a state register (the MPC), a microcode ROM, and
a $DECODE$ component. Several programs from Gabriel [9] and Henderson
[16] were run as simulations at the Mossim level. The largest test was the
compilation of the LispKit compiler from Henderson.

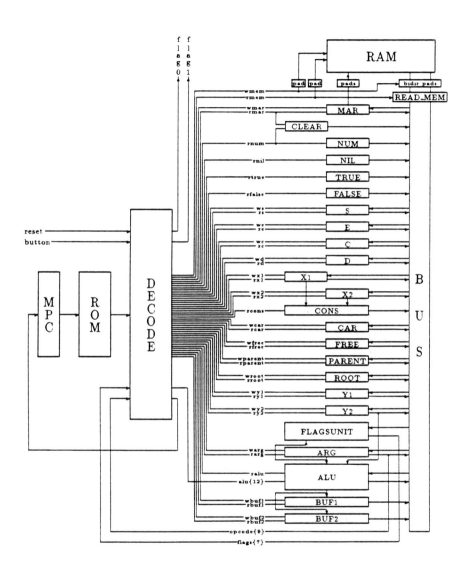

Figure 3.2: Register Transfer Level View of SECD Machine

3.4.4   SECD layout

The SECD chip has 4 major functional components: the **control unit**, the **datapath**, the **shift registers** and the **pad frame** as depicted in Figure 3.3.

1. The *control unit* interprets SECD machine instructions, breaking them up into a stream of micro-instructions to be effected one at a time by the *datapath* unit. It is conceived as a finite state machine whose state is held by a micro program counter (MPC) register which always refers to the current micro-control step. Inputs to the *control unit* include: external *reset* and *button* signals, status *flags*, and a 9–bit *opcode* (the code of the current machine instruction). Outputs include *read* and *write* signals for registers within the *datapath* and also for memory, *alu* control signals, the bi-directional pad control signal *bidir*, and the two state flags *flag0* and *flag1*.

2. The *shift registers* provide a means of entering test vectors and examining the state of the chip, giving us a passable ability to test the chip in operation. They also permit independent testing of the *datapath* and *control unit* components. Most signals passing between the *control unit* and the *datapath* are routed through the *shift registers*. These include *read* and *write* signals, *alu* signals, and status *flags*. For observability and controllability, it also routes some signals which are functionally internal to the *control unit* (specifically the 5 select signals, and the *mpc* register contents). Thus all these signals may be read and/or altered, though in normal operation, signals pass through uninterrupted. The *shift registers* use a clock input independent from the system clock.

3. The *datapath* executes simple micro-operations signalled by the *control unit*. The *datapath* unit is built as an ensemble of devices – registers, the arithmetic unit, and memory – communicating via a common *bus*. The operations performed by the *datapath* unit include: copying the value of a selected register onto a *bus*, storing a value from the *bus* into a selected register, the list manipulating functions cons, car, cdr, the arithmetic operations of addition, subtraction, decrementation, and setting status flags. Besides read and write lines and the arithmetic operator select lines, the only other input control line is the clock (Φ_A, which clocks the writing of registers). Outputs include the status *flags*, the memory address register (for the off chip RAM), the lower 9 bits of the *arg* register (the *opcode*) and the 32 bit *bus*.

4. The chip is framed by a set of *I/O pads* which connect the chip to the outside world. Bidirectional i/o pads connect the data bus to the

external memory. The default mode for the pads is input, switching to output mode only when writing to memory. A set of busgates isolates the input from the datapath bus unless a read memory operation is signalled. A 14 bit *mar* exported off chip from the datapath selects the memory location. The SECD chip is packaged in a 64-pin DIP. External (off chip) signals fall within three groups: signals used for normal operation of SECD, pwr/gnd, and SECD testing signals.

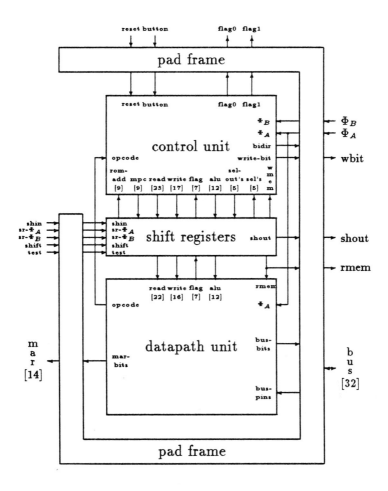

Figure 3.3: SECD Chip Major Subcomponents

3.5   SECD Formal Specification

Three levels of description of the SECD chip have been specified in HOL:
the **low level** consists of primitive gates (AND, OR, latch, etc.) and regular
structures (ROM's, PLA's), a **register transfer level** (rtl) model, consisting
of multi bit registers and functional components such as ALU's, etc. and the
top level specification, a representation of the abstract SECD architecture
defined earlier. We will focus on the rtl and top levels, limiting discussion
of the low level to a few key elements that impact the sequel. We adopt a
convention of subscripting the explicit time parameters to indicate the granularity of time associated with each level, low level (fine) grain time t_f, rtl
(medium) t_f, and top level (coarse) t_c.

3.5.1   The Low Level Definition

The low level definition uses simple gates as base components, and bears a
one-to-one relation to the layout of gates in the SECD chip. Specifying the
memory elements (latches) must be understood in relation to the granularity
of time associated with this level of description. We use the coarsest granularity of time that captures the behaviour of the clocks: using 4 points per
clock cycle. Constraints on clock behaviour require that the separate system
and shift register clock signals are not cycled simultaneously, and that when
started, the clocks always complete a full cycle *i.e.* the system clock cycle is
never interrupted by a series of shift register clock cycles, and vice-versa. All
latches on the SECD chip are level-triggered. Additionally, we require the
reset input be asserted when the system clock starts cycling, so that we have
a deterministic state on startup.

   The standard design regimen for 2-phase non-overlapping clocking, that of
linking memory devices clocked on opposite clock phases with combinational
logic, was violated in one part of the SECD chip design. Two registers clocked
on the same clock phase are wired in series, with the output of the first feeding
the input of the second through strictly combinational logic. This odd circuit
feature will degrade chip performance, but at sufficiently slow clock rates will
still function properly, since the output of level triggered latches change at
the start of the clock pulse, and no circular feedback path exists. This single
piece of questionable design affected the entire chip specification, and added
an obligation to prove the impossibility of circular feedback. A somewhat
nonstandard definition of the primitive latch was required to express this
behaviour appropriately, expressing present state as a function of the clock
signal at the same moment, instead of the previous moment. Notice the
difference between this definition and that of *del2* on page 136.

```
latch clk inp state =
  !t. state t = (clk t) => inp t | state (PRE t)
```

3.5.2 Register Transfer Level

The Register Transfer Level (rtl) model of the chip has a similar component hierarchy as the low level view, but does not have the same depth. It has 3 major components: the control unit, the datapath, and the pad frame (*CU, DP*, and *rt_PF* respectively). The *rt_PF* definition is the proven behaviour of the padframe component of the lower level. The *DP* is the conjunction of the same set of components as the low level view, but the behaviour of each component replaced the primitive definitions. The *CU* differed most from the lower level. The layout hierarchy was replaced by one that better reflected its functionality, using three components for its definition: a *state register*, a microcode *ROM*, and a *decode* section. Conjoining the definitions of these 3 parts gives a typical finite state machine behaviour, defining the next state and current output values as a function of the current state and current input values. Notably absent from this view is the shift register component. Under normal operation, these are intended to be transparent to the operation of the chip, and thus constraining their controls appropriately enables us to abstract it out of this level of description completely.

The external *memory* is represented only at the register transfer level, and the conjunction of the chip description and a simple memory with *Fetch* and *Store* commands constitutes the system at this level.

Temporal representation

The time granularity at this level corresponds to clock cycles. Selecting which point of the cycle to use for this time grain was complicated by the outputs of registers clocked on different clock phases feeding into a single register (*i.e.* this is the case when two registers clocked on the same phase are wired in series). The part of the clock cycle over which the output of the register is stable is determined by the clock phase used to clock the register. In order to have a uniform temporal abstraction for the time parameter for each input type, we selected the time that the first clock phase is asserted as the point at which the rtl state is defined.

Two different types of registers are used. The control unit uses pairs of latches clocked on opposite clock phases, while the datapath registers clock write operations on ϕ_B to prevent transitory spikes on state transitions from causing unwanted writes. The low level circuit and the rtl specification for each follow.

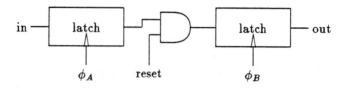

```
MPC9 clk in out =
   (out 0 = #000000000) ∧
   (!tₘ. out (tₘ+1) = clk tₘ => in tₘ | state tₘ)
```

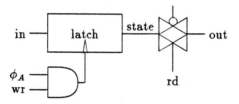

```
reg in out clk wr rd state =
(!tₘ. state tₘ =
        clk tₘ => (wr tₘ => in tₘ | state (PRE tₘ))
              | state (PRE tₘ)) ∧
(!tₘ. rd tₘ ==> (out tₘ = state tₘ))
```

In both definitions, the clock signal *clk* is an abstraction of the two distinct clock phases (ϕ_A and ϕ_B), and is asserted when the system clock cycles. Note that the *mpc* register specification does not include the *reset* input explicitly. However, constraints on the *reset* signal force an initialisation at power up, and this constraint and the signal appear as a determined value in the register at time $t_m = 0$. The typical datapath register has a gated output, and is clocked by both the system clock phase and the specific write signal. The clock signal was included here to permit eventual extension of the verification to the test mode of chip operation. The difference in the definition of the registers and the choice of point on the time cycle can be seen from the timing diagram in Figure 3.4.

The *reset* signal is held low at the start of the clock operation. The mpc register content at t_m+1 is a function of the inputs at time t_m. If the

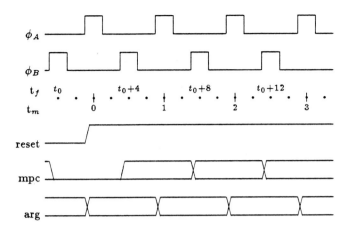

Figure 3.4: Relating low and rtl times

arg register (a typical datapath register) content at time t_m feeds into the first *mpc* latch at the same time, this relation holds, as it also does if the *mpc* content itself feeds back into this latch. Sampling one point earlier would have required distinct temporal abstractions for the signals arising from registers clocked on different clock phases.

The control unit

The state register component for the control unit includes not only the *mpc* register, but a 4-deep *stack* for microcode subroutine calls as well. The definition of these registers is altered slightly, with no provision for resetting. The clock signal fed to these latches is the same system clock, and the load signal is the OR of the *pop* and *push* control signals, so latching only occurs when a stack command is issued.

```
Stack_register9 Clocked load in_sig out_sig =
 !tm. out_sig (tm+1) =
       Clocked tm => (load tm => in_sig tm | out_sig tm)
                  | out_sig tm
```

The datapath

The datapath consists of the conjunction of the components shown in Figure 3.2. The bus is expressed as the wiring together of the components. All registers are instances of the previously shown definition. There are 3 elements that are worth a closer examination: the *FLAGSUNIT*, the *ALU*, and the *READ_MEM* component.

The *READ_MEM* component is a *busgate* (a set of transmission gates), which connect the bus to the bidirectional i/o pad output. This is required since the pads default to input operation, but only when a read memory instruction is asserted can the input be permitted to drive the bus. A *busgate* is defined as follows:

```
busgate in_val rd out =
     !t. (rd t) ==> (out t = in_val t)
```

The *ALU* can perform 9 distinct operations. Six of these are used exclusively in garbage collection: *replcar, replcdr, setbit31, setbit30, resetbit31*, and *resetbit30*. The remaining operations are the arithmetic operations: *add, sub*, and *dec*. The *mul, div*, and *rem* operations were not implemented, and selecting any of these defaults to the *dec* operation. The output of the *ALU* is gated onto the bus controlled by the *ralu* signal. If no *alu* operation is selected, the value computed is not significant, and thus we define this component by the use of implication, typically:

```
add t ==>
  (alu t = ADD28 (atom_bits(arg t)) (atom_bits(bus t)))
```

(The *ADD28* function is the specification for the operation performed by the low level component, and matches what is specified at the top level.) Just as the behaviour for no operation is of no concern, so the output when more than one operation is selected should be undefined. This is expressed in the specification by the use of a *one_asserted* predicate which implies the desired behaviour. This predicate states that if one of the signals is asserted, the remainder are not.

Finally, 6 status flags emerge from the *FLAGSUNIT* : *atomflag, bit30flag, bit31flag, zeroflag, eqflag*, and *leqflag*. The *bit* flags are used by the garbage collector. The *atomflag* indicates if the value on the bus is an atom record, as opposed to a cons cell record. The unary flag operations operate on the bus value, the binary use the content of the *arg* register in addition.

The padframe

The padframe definition includes input, output, and bidirectional pads. The input and output pads are defined by stating the equality of the pairs of nodes connected by the pads. The bidirectional pads are defined by:

```
(bidir t => (mem_bits t = bus_pins t)   % read memory %
        |  (bus_pins t = bus_bits t))  % write memory %
```

where *bus_pins* are the off-chip side, *bus_bits* are the bus nodes, and *mem_bits* are the nodes linking the pads to the input of the *READ_MEM* unit described earlier.

3.5.3   The Top Level Specification

At the most abstract level, the SECD machine is defined in terms of transformations to S-expressions in the 4 stacks, as shown previously in Table 3.1. A formal specification of the top level behaviour is ideally defined in terms of transformations to an S-expression data type that closely resembles this elegant definition. The closer the resemblance the better we are assured that the HOL specification is equivalent.

The method used by SECD for implementing recursive function definitions as closures with a circular environment component (see [16]) raises the complexity of the data representation problem considerably. Such circular S-expression lists, created by a destructive operation, cannot be mapped to a simple recursive data type. Further, structure is shared by S-expressions, particularly the environment component of closures. Mutually recursive function closures each reference the same environment, which is also in the *E* stack. When a destructive replace operation is performed (by executing a RAP instruction) to create the circular list structure, the change affects the common component of all closures simultaneously.

Thus, a much lower level of representation has been chosen to describe the top level specification. Rather than directly defining transformations on structures of an S-expression data type, we define an abstract memory type which can contain representations of S-expressions. Further, we define a set of primitive operations upon the memory which correspond to the operations on S-expressions, namely *cons, car, cdr, atom, replcar, dec, eq, leq, add,* and *sub.* The 4 state registers contain values that reference the appropriate S-expression representation. Finally, an additional *free* register containing a value to access the free list structure is needed to define the *cons* operation. The state of the machine is then defined by a tuple:

$$(S, E, C, D, Free, memory, FSM\_state)$$

where the FSM_state is one of the 4 major states of the top level finite state machine view of the machine (as in Figure 3.1).

The abstract memory type μ is basically a function: $\mu = \delta \rightarrow (\delta \times \delta \cup \alpha)$ where δ is the domain of the function and α is the set of atoms: $\alpha = integers \cup symbols$. The set of symbols includes the symbolic constants: *T*, *F*, and *NIL*. The domain of the function is chosen to be the type of 14 bit words (*word14*, a 14 bit instance of the general *wordn* type of fixed width bus values), matching the type that is used by the implementation definition. We extend the definition of memory to incorporate garbage collection features, by adding *mark* and *field* bits to each cell:

$$\mu = \delta \rightarrow ((bool \times bool) \times (\delta \times \delta \cup \alpha))$$

Additionally, we include the *replcdr, setf,* and *setm* operators used by the garbage collector, as well as a *Garbage_collect* function, which is left undefined for the first proof attempt. Extractor functions *mark, field, Int_of,* and *Atom_of* are provided for the values returned by the μ function. The relevant built-in functions and their types are summarized in Table 3.2.

Operation	Type
Car, Cdr	$(\delta \times \mu \times \delta) \rightarrow (\delta)$
CAR, CDR	$(\delta \times \mu \times \delta) \rightarrow (\delta \times \mu \times \delta)$
Cons, Cons_tr	$(\delta \times \delta \times \mu \times \delta) \rightarrow (\delta \times \mu \times \delta)$
EQ, LEQ	$(\delta \times \delta \times \mu \times \delta) \rightarrow bool$
ADD, SUB, MUL, DIV, REM	$(\delta \times \delta \times \mu \times \delta) \rightarrow (\delta \times \mu \times \delta)$
Replcar, Replcdr	$(\delta \times \delta \times \mu \times \delta) \rightarrow (\delta \times \mu \times \delta)$
setm, setf	$(bool \times \delta \times \mu \times \delta) \rightarrow (\delta \times \mu \times \delta)$
mark, field	$\delta \rightarrow \mu \rightarrow bool$
Int_of	$(\delta \times \mu \times \delta) \rightarrow integer$
Atom_of	$(\delta \times \mu \times \delta) \rightarrow \alpha$
Atom	$(\delta \times \mu \times \delta) \rightarrow bool$
Is_int	$(\delta \times \mu \times \delta) \rightarrow bool$
Is_TRUE	$(\delta \times \mu \times \delta) \rightarrow bool$
Garbage_collect	$(\mu \times \delta) \rightarrow (\mu \times \delta)$

Table 3.2: Primitive Operations on Abstract Memory Data Type

As seen in the table, many of the functions return triples, consisting of a memory cell, a memory, and a second cell which represents a free list pointer. Operations such as *Car, Cdr, EQ, LEQ*, etc. do not alter memory, while *Cons,*

ADD, SUB, setm, setf, Replcar, Replcdr do alter one cell in the memory, and thus must return the new memory. In order to permit composition of the primitive memory operations, we provide the *CAR* and *CDR* functions which return the unaltered memory and free pointer. For example, to access an argument to a *LD* command, we can write the following:

$$let\ m = Int\_of(CAR(CAR(CDR(c, MEM, free)))).$$

```
LDF_trans (s:δ,e:δ,c:δ,d:δ,free:δ,MEM:μ) =
  let cell_mem_free =
    M_Cons_tr(s, M_Cons_tr(e, M_CAR(M_CDR(c,MEM,free))))
  in
  (cell_of cell_mem_free,                          % s %
   e,                                              % e %
   M_Cdr(M_CDR(c, mem_free_of cell_mem_free)),     % c %
   d,                                              % d %
   free_of cell_mem_free,                          % free %
   mem_of cell_mem_free,                           % memory %
   top_of_cycle)                                   % state %
```

Table 3.3: Transition for LDF Instruction

In Table 3.3 we provide the top level transition specification for the *LDF* instruction. Following this, in Table 3.4 the set of 21 such transitions are used to define the next state of the machine for each instruction, as well as the top level specification for the SECD. The top level function, *SYS_spec*, closely resembles the top level finite state machine of Figure 3.1, with 4 states, 2 possible transitions from 3 of them, and 18 transitions from the *top_of_cycle* state, one for each (implemented) machine instruction. The system must be assured of starting out in the *idle* state. The given types of all system parameters have the subscript "$_c$". This identifies them as signals sampled at a *coarse* grain of time.

The type μ was defined in HOL as the abstract data type *mfsexp_mem*, and the type α was defined as the type *atom*. Both types have associated *REP* and *ABS* functions to map between the abstract and representing types. For example, the term *"REP_mfsexp_mem(memory:(word14,atom)mfsexp_mem)"* has the type:

$$word14 \rightarrow ((bool \times bool) \times (word14 \times word14 \cup atom)).$$

```
NEXT (s:δ,e:δ,c:δ,d:δ,free:δ,MEM:μ) =
 let instr = M_int_of(M_CAR(c,MEM,free))
 in
 ((instr = LD)   =>    (LD_trans   (s,e,c,d,free,MEM)) |
  (instr = LDC)  =>    (LDC_trans  (s,e,c,d,free,MEM)) |
  (instr = LDF)  =>    (LDF_trans  (s,e,c,d,free,MEM)) |
     ...
  (instr = LEQ)  =>    (LEQ_trans  (s,e,c,d,free,MEM)) |
% (instr = STOP) %    (STOP_trans (s,e,c,d,free,MEM))))
```

```
SYS_spec (s:δc) (e:δc) (c:δc) (d:δc)
         (free:δc) (MEM:μc)
         (button:boolc) (state:statec) =
 (state 0c = idle) ∧
 ! tc.
 ((s(tc+1), e(tc+1), c(tc+1), d(tc+1),
   free(tc+1), MEM(tc+1), state(tc+1)) =
 ((state tc = idle)
  => (button_pin tc
     => (M_Cdr(M_CAR(NUM_addr, MEM tc, free tc)),
         NIL_addr,
         M_Car(M_CDR(NUM_addr, MEM tc, free tc)),
         NIL_addr, NIL_addr, MEM tc, top_of_cycle)
      | (s tc, e tc, c tc, d tc, free tc, MEM tc, idle))
 |(state tc = error0)
  => (button_pin tc
     => (s tc, e tc, c tc, d tc, free tc, MEM tc, error1)
      | (s tc, e tc, c tc, d tc, free tc, MEM tc, error0))
 |(state tc = error1)
  => (button_pin tc
     => (s tc, e tc, c tc, d tc, free tc, MEM tc, error1)
      | (s tc, e tc, c tc, d tc, free tc, MEM tc, idle))
 |% (state tc = top_of_cycle) %
   (NEXT (s tc, e tc, c tc, d tc, free tc, MEM tc)))
```

Table 3.4: Top Level Specification

Memory Abstraction

Abstracting from the implementation memory to the abstract memory type maintains the mark and field bits unchanged, and maps the 28 bit field to the appropriate *cons*, *integer*, or *symbol* record based on the record type bits. The memory abstraction function is defined in two steps: first a function maps records in the range of the implementation memory into the range of the representative type for abstract memories, and then *ABS_mfsexp_mem* is applied to this function composed with the implementation memory:

```
Mem_Range_Abs:word32->((bool#bool)#((word14#word14)+atom))w =
  ((mark_bit w),(field_bit w)),
  ((is_symbol w) => INR(Symb(Val(Bits28(atom_bits w)))))
  |(is_number w) => INR(Int(iVal(Bits28(atom_bits w)))))
                    | INL(car_bits w, cdr_bits w))

mem_abs (M:word14->word32) = ABS_mfsexp_mem(Mem_Range_Abs o M)
```

The Mem_Range_Abs function is total, and the unused record type (#01) gets mapped to the *cons* type record of the abstract memory. This ensures that the result of composing it with an implementation memory always returns an object in the representative type of the abstract memory, and thus applying *REP_mfsexp_mem* to *mem_abs M* at some value v will be *Mem_Range_Abs (M v)*. It further ensures the desirable property that every non-atomic record is a *cons* type record.

Temporal Abstraction

The coarsest grain of time used to describe the system corresponds to the points when the system is in major states of the top-level FSM (*Idle, Error0, Error1,* and *Top_of_Cycle*). We map from this coarse granularity to the medium grain points of time when specific microcode addresses are in the *mpc* register. The mapping is not a linear function, since the number of cycles needed to execute any machine instruction varies, and can vary between executions of the same instruction. The latter differences arise due to garbage collection calls during instruction execution, as well as varying search distances required to load values from the environment. The definition of the required mapping functions is well described in [25].

3.6   Verification of the SECD Design

Within the accuracy of the model of the lowest level of description, and subject to given constraints, a proof of correctness will show that a lower level description ensures the behaviour of the higher level of description. Parameters to the higher level description are abstractions of the lower level ones, with regard to temporal granularity and data type. Thus, we can view each level description as a specification, and also as an implementation of the higher level specification. Using this terminology, the goal of a proof generally looks like:

constraints ⊃
  implementation (state) (inputs) (outputs) ⊃
    specification (abs o state) (abs o inputs) (abs o outputs)

For the register transfer to top level specification proof, the abstraction must map between two granularities of time, as well as different data types, particularly for the memories. The register transfer definition includes a memory function with simple Fetch and Store operations only, while the top level uses an abstract memory data type. The task of the verification is to show that the sequence of operations performed at the register transfer level commutes with the specification transition at the abstracted top level.

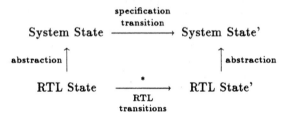

A complete verification of the SECD system from lowest level description to top level specification requires correctness theorems relating the low and register transfer levels, and one relating the register transfer and top levels. The first of these will not be discussed in this paper.

This section of the paper addresses the constraints on the scope of the proof, the form of the correctness goal, and the approach used in achieving the proof. Each major step in the proof is outlined, using the LDF transition as the running example.

3.6.1   Constraints

Four major constraints limit the scope of proof:

clock_constraint We restrict the proof to normal operating mode (*i.e.* the shift registers are not operating).

```
clock_constraint SYS_Clocked = ! t_m. SYS_Clocked t_m
```

reserved_words_constraint There are 3 reserved words in memory that contain the symbolic constants for *NIL, T,* and *F.*

```
reserved_words_constraint mpc memory =
 !t_m. (state_abs (mpc t_m) = top_of_cycle) ==>
  ((memory t_m NIL_addr =
    bus32_symb_append #00000000000000000000000000000000) ∧
   (memory t_m T_addr   =
    bus32_symb_append #00000000000000000000000000000001) ∧
   (memory t_m F_addr   =
    bus32_symb_append #00000000000000000000000000000010))
```

well_formed_free_list · Informally, we want the free list to be a *cdr* linked list, containing only *cons* cells, aside from the last cell which is *NIL.* Only cells that are not used in any data structure that is part of the computation (*i.e.* cells accessible from the *s, e, c,* or *d* registers), and cells that are not reserved words should appear in the free list. Finally, cells should only appear in the list at most once. For the formal specification, we limit the constraint to the maximum number of *cons* operations in any SECD instruction: the AP instruction performs the operation 4 times. The requirement that the cells be *cons* records ensures the symbolic constants are not in the list. The function *nth* applies its 2nd argument *n* times to the last argument.

free_list_constraint, derived from this and the previous constraint, simply asserts that the first 4 pointers of the free list are not *NIL_addr,* and is used in the microprogramming level proofs.

```
well_formed_free_list mpc free memory =
 ! t_m. (state_abs (mpc t_m) = top_of_cycle) ==>
 ! n. (n < 4) ==>
  ((is_cons (nth n ((memory t_m) o cdr_bits)
                   (memory t_m (free t_m)))) ∧
   (~(nth n(cdr_bits o (memory t_m))(free t_m) = NUM_addr)) ∧
   !n'. (n' < n) ==>
    ~(nth n (cdr_bits o (memory t_m))(free t_m) =
      nth n'(cdr_bits o (memory t_m))(free t_m)))
```

valid_program_constraint The state of the machine must pattern match
with the left side of the abstract machine transition for one of the 18
implemented machine instructions, or have a problem loaded in memory
when the machine is directed to start computation. The constraint gives
a quite detailed specification of the type and in some cases the value of
each record in memory involved in the pattern match. A less detailed
constraint, limiting the possible machine instructions and requiring the
arguments to the LD instruction represent non-negative values, was sufficient for the proof of liveness of the machine. It was defined independently and proved to be implied by the more detailed constraint shown
here.

This valid program constraint must be assured by the correctness of
the compilation process. As with the *free_list_constraint* we restrict the
constraint to times after the machine is initialised.

```
valid_program_constraint memory mpc button_pin s e c d =
 !t_m .
   (((state_abs(mpc t_m) = idle) ∧ button_pin t_m) ==>
     (is_cons(memory t_m NUM_addr)) ∧
     (is_cons(memory t_m(car_bits(memory t_m NUM_addr)))))) ∧
   ((state_abs(mpc t_m) = top_of_cycle) ==>
     let head_c = memory t_m(c t_m)
     in
     ((is_cons head_c) ∧
      let instr' = memory t_m(car_bits head_c)
      and next_c = cdr_bits head_c
      in
      ((is_number instr') ∧
       let instr = atom_bits instr'
       in
        (  ...
          ((instr = ^LDC_instr28) ∧
           (is_cons(memory t_m next_c)) ∧
           (is_atom(memory t_m(car_bits(memory t_m next_c)))))
          ) ∨
          ((instr = ^LDF_instr28) ∧
           (is_cons(memory t_m next_c))
          ) ∨
           ...
          ((instr = ^STOP_instr28) ∧
           (is_cons(memory t_m(s t_m))) ∧
           (next_c = NIL_addr))))))
```

3.6.2 Structure of the proof

The goal for this level of the proof is shown in Figure 3.5. Most of the signals are simply abstracted from the medium to the coarse time granularity. Additionally, the memory and state are abstracted to the appropriate data type as well. The validity of the temporal abstraction of the one input signal which is an argument to the top level specification (*i.e.* the *button_pin* input) should be carefully considered. We can show that this input affects the next state of the machine at the medium time granularity in only 3 places in the microcode, and these 3 places correspond to points in coarse grain time, since they are *major states*.

```
"((clock_constraint SYS_Clocked)                     ∧
  (reserved_words_constraint mpc memory)             ∧
  (well_formed_free_list mpc free memory)            ∧
  (valid_program_constraint memory mpc button-pin s e c d))   ==>
 (SYS memory SYS_Clocked
      mpc s0   s1   s2   s3
      button_pin
      flag0_pin    flag1_pin   write_bit_pin   rmem_pin
      bus_pins     mar_pins
      s   e   c   d   free
      x1  x2  y1  y2   car   root   parent
      buf1     buf2     arg)                         ==>
 SYS_spec   ((mem_abs o memory)   when (is_major_state mpc))
            (s                    when (is_major_state mpc))
            (e                    when (is_major_state mpc))
            (c                    when (is_major_state mpc))
            (d                    when (is_major_state mpc))
            (free                 when (is_major_state mpc))
            (button_pin           when (is_major_state mpc))
            ((state_abs o mpc)    when (is_major_state mpc))"
```

Figure 3.5: RTL ⊃ top level goal

Although the lower (rtl) level definition of the system for this proof consists of the wiring together of high-level components defined behaviourally, it is not feasible to undertake the proof of the goal in Figure 3.5 directly. The size of terms generated alone makes this impossible to manage. The proof is instead undertaken in several stages.

First, a simplified, flattened specification for the register transfer level definition was obtained. Second, this specification was used to derive theorems for the change of state effected by each of the 400 microcode instructions. These theorems represent the "Phase level" of description of Anceau [1]. Third, a sort of microcode level simulation used the 400 theorems to step through the microcode to produce theorems that summarized the behaviour of instruction sequences. This corresponds to the "microprogramming level" of Anceau. Fourth, these theorems were then used to prove a liveness property for the system. And last, we proved that the state transitions resulting from the register transfer level definition correspond to those defined at the top level. A description of each stage follows.

Unwinding the System Definition

The register transfer level of description consists of a conjunction of behavioural descriptions of major components. For example, the control unit consists of a *state register* (with 5 fields), a *ROM* and a *decode* unit. A specification for the conjunction of these three subcomponents is not appreciably simpler than the conjunction of the component behaviours individually. It is only when the whole system is assembled that simplifications become possible. A good example is the definition of the *ALU* subcomponent of the datapath. Its behavioural definition is constrained by a one-assertedness property on the control lines. This constraint is eliminated in the whole system because it is proved that the control unit assures this property of the control signals to the *ALU*.

Thus the normal approach, that of first defining a specification for each level of the hierarchy and then using each of these in turn to prove the next level of specification, was abandoned. Instead we simplified each level by UNWINDing existentially quantified variables where possible. This was carried out up to the top of the hierarchy where more substantial simplifications could be made. The process of creating these proofs was relatively straightforward, complicated only by the relatively large size of terms involved. However, simple rewrites could exhaust the Lisp namestack, so that the proof process had to be managed extremely closely (generally using quite low level tactics and rules).

Theorems for simplifying the control unit and datapath were undertaken first. These were achieved by rewriting with all subcomponent definitions, and UNWINDing and PRUNEing existentially quantified variables (hidden wires). The *one_asserted* constraint in the datapath was moved to the outermost level of the definition. Also, as all subcomponents have a quantified time parameter, the distinct quantified variables were replaced by a single quantified variable, encompassing the entire definition.

A theorem showing that the alu control outputs from the control unit have a *one_asserted* property was proved. The intensive computation required to prove inequality of two values of a specified wordn type made it desirable to prove an exhaustive set of theorems for all possible values of each subfield of the microcode ROM output. Rather than repeatedly proving the cases for each of the 400 ROM addresses, this required 68 theorems all told. These theorems were of the form:

```
Alu_base_0001 =
⊢ (Alu_field(ROM_fun(mpc t_m)) = #0001) ==>
   ((Alu_field(ROM_fun(mpc t_m)) = #0001) = T) ∧
   ((Alu_field(ROM_fun(mpc t_m)) = #0010) = F) ∧
   ...
   ((Alu_field(ROM_fun(mpc t_m)) = #1011) = F) ∧
   ((Alu_field(ROM_fun(mpc t_m)) = #1100) = F)
```

In addition, theorems for the value of the Inc9 function, used to calculate the next mpc address in sequential microcode sequences, for each of 400 addresses were proved.

```
⊢ Inc9 #000000000 = #000000001
```

At the top of the hierarchy, the static RAM definition of memory was added in, and the CU and DP simplifications were applied to create one flattened expression for the SYS implementation. The *one_asserted* constraint fell out. All internal read and write control lines were eliminated. The only remaining existentially quantified variables, *bus_bits*, *mem_bits*, and the *alu*, were quantified over all conjuncts. The most important step involved moving the time parameter outside the existentially quantified internal lines, and replacing these time varying lines with static values. The equivalence of this transformation was proved, and specialised conversions designed. This key step permits evaluating the entire expression when constraining some value at a given time t_m; in fact we evaluated the simplified expression for the system description under the constraint $mpc\ t_m = x$, where x is one of the 400 microcode addresses. From this stage onward, equivalence was no longer attempted, so the final form of the theorem has the clock_constraint and the

```
Base_thm =
["clock_constraint SYS_Clocked"; ^SYS_imp]
⊢ ? bus_bits_t mem_bits_t alu_t.
      (mpc(SUC t_m) =
       (((Test_field(ROM_fun(mpc t_m)) = #0001)    ∨
         (Test_field(ROM_fun(mpc t_m)) = #0011) ∧
          field_bit bus_bits_t                                ∨
          ...
         (Test_field(ROM_fun(mpc t_m)) = #1011)) =>
         A_field(ROM_fun(mpc t_m)) |
         ((Test_field(ROM_fun(mpc t_m)) = #1100) =>
          s0 t_m |
          ((Test_field(ROM_fun(mpc t_m)) = #0010) =>
           Opcode arg t_m | Inc9(mpc t_m))))) ∧
       ...
      (memory(SUC t) =
       ((Write_field(ROM_fun(mpc t)) = #00001) =>
        Store14(mar_pins t)(bus_pins t)(memory t) |
        memory t)) ∧
      ((Read_field(ROM_fun(mpc t_m)) = #00010) ==>
       (bus_bits_t = mem_bits_t)) ∧
      ...
      ((Alu_field(ROM_fun(mpc t)) = #0001) ==>
       (alu_t = DEC28(atom_bits(arg t)))) ∧
      ...
      (~(Write_field(ROM_fun(mpc t_m)) = #00001) ∧
       (Read_field(ROM_fun(mpc t_m)) = #00010) ==>
       (bus_pins t_m = memory t_m(mar_pins t_m))) ∧
      ...
      (s t_m = ((Write_field(ROM_fun(mpc t_m)) = #00110) =>
               cdr_bits bus_bits_t | s(PRE t_m))) ∧
      ...
      (flag1_pin t_m = (mpc t_m = #000101011) ∨
                       (mpc t_m = #000011000))
```

Figure 3.6: Base_thm: the rtl definition simplified

SYS definition[6] as assumptions, and the conclusion is the simplified expression for the system, with the 3 existentially quantified variables. Also the conjunct giving original value of the mpc (*i.e.* at time $t_m = 0$) was dropped from the

---

[6] We will abbreviate this constraint as SYS_imp in the following material.

expression. Representative samples of each type of conjunct of the theorem are shown in Figure 3.6.

Phase level: effect of each microinstruction

The simplified expression for the system was next utilised to produce theorems for the state of the system at each microcode address. A sample theorem for the system when the value #001100001 is in the *mpc* register is shown in figure 3.7. This is the first instruction of the microcode sequence for the LDF instruction, and effects a transfer of the content of the *e* register to the *x2* register. The two dots represent the clock constraint and the system implementation assumptions. The large number (400) of such theorems required that they be generated without user intervention, and as efficiently as possible. A single proof function was designed for this purpose, and while efficiency was important in its design, the required generality was a limiting factor. Generating the set of theorems took approximately 36 hours running on a dedicated Sun 3/60 workstation with 16 megabyte memory. The theorems were generated in groups that corresponded to sets of SECD instruction code sequences or subroutines, to save search times for theorems at the next proof level.

The proof function works as follows:

1. Fetch the appropriate microcode lemma for the given address, and specialise it with the specific mpc value. This lemma has the form:
   . ⊢ ROM_fun(mpc t_m) = #000000000000000000110001000
   The dot to the left of the turnstile symbol represents the assumption *mpc t_m = #001100001*, in this and all the following theorems.

2. Substitute for the 27 bit constant with the 27 bit Bus, then for each field, apply the appropriate field selector function, rewrite with the definition of the function, and return the Bus form of the field value to the wordn constant form. This theorem is then resolved with the CU_proofs theorem for that field value of constant, and split into conjuncts, giving a set of theorems each having the form:
   . ⊢ (Test_field(ROM_fun(mpc t_m)) = #0101) = F
   This is done for each required value for the Test, Read, Write, and Alu fields of the 27 bit value.

3. The relevant Inc9 theorem is retrieved from Inc9_proofs and specialised:
   . ⊢ Inc9(mpc t_m) = #001100010.

4. An Afield theorem is obtained similarly to the other field theorems:
   . ⊢ A_field(ROM_fun(mpc t_m)) = #000000000.

```
SYS_lemma_97 =
 .. ⊢ (mpc t = #001100001) ==>
        (mpc(SUC t) = #001100010) ∧
        (s0(SUC t) = s0 t) ∧
        (s1(SUC t) = s1 t) ∧
        (s2(SUC t) = s2 t) ∧
        (s3(SUC t) = s3 t) ∧
        (memory(SUC t) = memory t) ∧
        (x2 t = e(PRE t)) ∧
        (rmem_pin t = F) ∧
        (buf1 t = buf1(PRE t)) ∧
        (buf2 t = buf2(PRE t)) ∧
        (mar_pins t = mar_pins(PRE t)) ∧
        (s t = s(PRE t)) ∧
        (e t = e(PRE t)) ∧
        (c t = c(PRE t)) ∧
        (d t = d(PRE t)) ∧
        (free t = free(PRE t)) ∧
        (x1 t = x1(PRE t)) ∧
        (car t = car(PRE t)) ∧
        (arg t = arg(PRE t)) ∧
        (parent t = parent(PRE t)) ∧
        (root t = root(PRE t)) ∧
        (y1 t = y1(PRE t)) ∧
        (y2 t = y2(PRE t)) ∧
        (write_bit_pin t = T) ∧
        (flag0_pin t = F) ∧
        (flag1_pin t = F)
```

Figure 3.7: Theorem for microcode instruction at address 97

5. The flag output expressions are evaluated with the mpc value substituted in, to generate a theorem of the form:
   . ⊢ (mpc t_m= #000010110) ∨(mpc t_m= #000011000) = F.

6. The set of all the proceeding theorems is assembled as a list, and substituted into the Base_thm, using a template, and a series of rewrites reduces constant boolean expressions and removes eliminated existentially quantified variables, followed by a pruning of any remaining existentially quantified variables.

7. If any existentially quantified variables remain, they are eliminated by

a specialised rule that uses theorems about the existence of fields of an existentially quantified variable:

$$\text{car\_bits\_thm} = \vdash \ ! \ y. \ ? \ x. \ \text{car\_bits} \ x = y$$
$$\text{cdr\_bits\_thm} = \vdash \ ! \ y. \ ? \ x. \ \text{cdr\_bits} \ x = y$$

Microprogramming level: symbolic execution

Just as every one of the 400 lemmas about the rtl description give us the next state in terms of the present state, we next want to combine these results to give the state after a number of steps in terms of the present state.

The definition for *SYS_spec* describes a 4 state fsm, with 2 transitions from each of 3 states controlled by the button input, and 18 possible transitions, one for each machine instruction code, from the 4th state. Further, 4 instruction sequences have a branch conditional upon some function of the state, and one (the sequence for the *LD* instruction) contains 2 loops. Thus there are minimally 28 paths to consider. Additionally, several instruction sequences call subroutines, and subroutine calls are nested. Under the *free_list_constraint*, the subroutines have a deterministic execution time.

The proof needed to show the state of the machine when it next reached a major state. This was easily achieved by a combination of $MATCH\_MP^{7}$ with the conjunct giving the value of $mpc(SUC \ t_m)$, and rewriting with all the other clauses of the *SYS_lemma_** theorems. Additionally, it was necessary to prove whether the system was in a major state at each step. A proof function was designed to produce such a pair of theorems, and a higher level function could call it recursively to produce a pair of theorems covering a sequence of steps. The subroutine correctness proofs were produced using these functions, plus another higher level function that could use a preproved segment proof pair (required when a subroutine call was involved).

The sequence for a machine execution, in this case the *LDF* instruction, is summarized in the two theorems in Figure 3.8. There are 5 assumptions, including:

1. "clock_constraint SYS_Clocked"

2. SYS_imp

3. "free_list_constraint mpc free"

4. "mpc t = #000101011"

5. "opcode_bits(memory t(car_bits(memory t(c t)))) = #000000011"

```
LDF_state =
.....  ⊢   (mpc(t_m+26) = #000101011) ∧
           (s0(t_m+26) = s0 t_m) ∧ (s1(t_m+26) = s1 t_m) ∧
           (s2(t_m+26) = s2 t_m) ∧ (s3(t_m+26) = #000000000) ∧
           (memory(t_m+26) = ^memory_exp) ∧
           (mar_pins(t_m+26) =
            cdr_bits(^memory_exp(cdr_bits(^memory_exp(c t_m))))) ∧
           (rmem_pin(t_m+26) = F) ∧
           (buf1(t_m+26) = buf1 t_m) ∧
           (buf2(t_m+26) = buf2 t_m) ∧
           (s(t_m+26) = cdr_bits(memory t_m(free t_m))) ∧
           (e(t_m+26) = e t_m) ∧
           (c(t_m+26) =
            cdr_bits(^memory_exp(cdr_bits(^memory_exp(c t_m))))) ∧
           (d(t_m+26) = d t_m) ∧
           (free(t_m+26) =
            cdr_bits(^memory_exp(cdr_bits(memory t_m(free t_m))))) ∧
           (x1(t_m+26) = free t_m) ∧ (x2(t_m+26) = s t_m) ∧
           (car(t_m+26) =
            car_bits(memory t_m(cdr_bits(memory t_m(c t_m))))) ∧
           (arg(t_m+26) = memory t_m(car_bits(memory t_m(c t_m)))) ∧
           (parent(t_m+26) = parent t_m) ∧ (root(t_m+26) = root t_m) ∧
           (y1(t_m+26) = y1 t_m) ∧ (y2(t_m+26) = y2 t_m) ∧
           (write_bit_pin(t_m+26) = T) ∧
           (flag0_pin(t_m+26) = F) ∧ (flag1_pin(t_m+26) = T)

where memory_exp =
 "Store14
   (cdr_bits(memory t_m(free t_m)))
   (bus32_cons_append #00 RT_CONS(free t_m)(s t_m))
   (Store14
    (free t_m)
    (bus32_cons_append
       #00 RT_CONS
       (car_bits(memory t_m(cdr_bits(memory t_m(c t_m)))))
       (e t_m))
    (memory t_m))"

LDF_Next = .....  ⊢ Next t_m(t_m+26)(is_major_state mpc)
```

Figure 3.8: Microprogramming level theorems for LDF instruction

The rather complex expression for the state of the machine becomes more comprehensible when only the values abstracted to the top level are considered. The *mpc* content corresponds to the *top_of_cycle* state. The *memory* has had 2 cells altered: the first cell has pointers to the code argument to the *LDF* instruction, and the current environment *e*. The second cell has pointers to the previous one, and the original stack *s*. The *s* register is updated to point to the latter cell. The *c* register now points to the rest of the control list after the *LDF* instruction and its argument. Notice that the *cdr* operation is performed on different memories: the first on the original memory, and the second after the memory is updated with the 2 rewritten cells. The *free* register has similarly been updated to the the 3rd cell in the original free list. The content of the other registers is either unchanged or irrelevant to the abstracted state. The expression for the updated memory appears several times, so has been extracted to simplify reading.

Liveness

If the SECD system initially reaches a major state, and every time it is in a major state all possible paths eventually return it to a major state, then the temporal abstraction function from the coarse granularity of time to the medium granularity is total. This important property is essential for the last part of the proof.

```
⊢ Next t1 t2 f = (t1 < t2) ∧ (f t2)  ∧
                 ! t.(t1 < t) ∧ (t < t2) ==> ∼ f t

⊢ (IsTimeOf      0  f t = f t ∧ !t'.(t'<t) ==> ∼ f t') ∧
  (IsTimeOf (SUC n) f t = ? t'. IsTimeOf n f t' ∧ Next t' t f)

⊢ TimeOf f n = @t.IsTimeOf n f t

⊢ when (s:num->*) (p:num->bool) = λn.s (TimeOf p n)

⊢ Inf f = ! t.? t'.(t<t') ∧ (f t')
```

Figure 3.9: Temporal Abstraction Function Definitions

The *state_abs* function is well-defined only when one of 4 values is in the

---

[7] *MATCH_MP* is a HOL variant of Modus Ponens.

mpc register, and these are precisely the 4 values for which the temporal abstraction predicate *is_major_state mpc* is true. Unfortunately, one cannot prove that the abstraction predicate holds at the time abstracted to by the predicate, *i.e. (is_major_state mpc) when (is_major_state mpc))t_c*, or in general *(P when P) t* (which is definitionally equivalent to *P(TimeOf P t)*). This limitation arises from the use of the Select operator @ in defining the temporal abstraction function TimeOf[8], given above. We overcome this obstacle by by proving a "liveness" property for the predicate used for the abstraction (*is_major_state mpc*). Liveness as defined by *Inf* states that the predicate is true infinitely often. The proof of this property is simplified by using the theorem *Inf_thm*:

```
⊢ ! f.(? t'ₘ. 0 < t'ₘ ∧ f t'ₘ) ∧
    (! tₘ. f tₘ ==> (? t'ₘ. tₘ < t'ₘ ∧ f t'ₘ)) ==> Inf f
```

which limits the proof requirement to showing that a major state is reached initially, and every time the machine is in a major state it will reach a major state again in the future. This limits the number of starting points to four, instead of every possible machine state (consisting of 400+ values that can be in the *mpc*). The microprogramming level theorems defining the *Next* times the temporal abstraction function holds are used for each branch of the proof.

Relating the Levels over Abstraction

The top goal splits into two parts: the state of the mpc when the machine first reaches a major state, and the state of the machine when it is next in a major state, in terms of its state in the previous major state (*i.e.* at time t_c). The first goal is quite simply solved by using SYS_lemma_0 to show that at time $SUC\ 0_m$ we reach a major state, and that at all times before that (namely time $t_m = 0$), the system is not in a major state. The required theorem has the form:

```
⊢ clock_constraint SYS_Clocked ==>
  ^SYS_imp ==>
  (((state_abs o mpc) when (is_major_state mpc))0_c = idle)
```

---

[8]From the defining axiom for the Select operator SELECT_AX: ⊢∀(P:* -> bool) (x:*). P x ==> P($@ P), it is necessary to show that P holds at some value x in order to derive that P($@ P) holds. See [15] for further description.

```
car_cdr_mem_abs_lemma =
⊢ is_cons(memory v) ==>
  !x. (M_Car(v,mem_abs memory, x) = car_bits (memory v)) ∧
      (M_Cdr(v,mem_abs memory, x) = cdr_bits (memory v))

number_mem_abs_lemma =
⊢ is_number(memory v) ==>
  !x. M_int_of(v,mem_abs memory, x) =
      iVal(Bits28(atom_bits (memory v)))

opcode_mem_abs_lemma =
⊢ is_cons(memory v) ==>
  is_number(memory(car_bits(memory v))) ==>
  !x. M_int_of(M_CAR(v,mem_abs memory,x)) =
      iVal(Bits28(atom_bits(memory(car_bits(memory v)))))

cons_mem_abs_lemma =
⊢ is_cons (memory free) ==>
  !v w.
  M_Cons(v,w,mem_abs memory,free) =
  (free,
   mem_abs(Store14 free (bus32_cons_append #00 RT_CONS v w) memory),
   cdr_bits (memory free))

Store14_cons_mem_abs_lemma =
⊢ (~(free = (cdr_bits(memory free))))        ==>
  is_cons (memory (cdr_bits (memory free))) ==>
  !v w z.
  M_Cons(v,w,
          mem_abs(Store14 free z memory),
  cdr_bits(memory free)) =
  (cdr_bits(memory free),
    mem_abs(Store14 (cdr_bits(memory free))
                    (bus32_cons_append #00 RT_CONS v w)
                    (Store14 free z memory)),
    cdr_bits (Store14 free z memory (cdr_bits (memory free)))))
```

Figure 3.10: Abstract Memory Theorems

We may split the second goal into a set of subgoals, corresponding to each transition defined in the SYS_spec. The transitions are determined by the state at time t_c, which is a function of the *mpc* at *TimeOf(is_major_state mpc)t_c*. Using the liveness property described previously, with the theorem TimeOf_TRUE:

$$\text{TimeOf\_TRUE} = \vdash \forall f \cdot \textit{Inf } f \implies (\forall n \cdot f \ (\textit{TimeOf } f \ n)),$$

to obtain the result that *is_major_state mpc* is true at all points of the coarser granularity of time. With this result, the state abstraction function is well defined, and the *mpc* has one of four values at all points in the coarser granularity of time.

The specification for each transition defines the new values for each of the *s, e, c, d*, and *free* registers and the abstract memory in terms of abstract memory operations on the previous values of the registers and abstract memory. A set of theorems relating abstract memory operations to rtl level operations is given in Figure 3.10. The theorem *opcode_mem_abs_lemma* combines the two previous theorem results for the machine instruction code value evaluation. The last theorem, *Store14_cons_mem_abs_lemma*, is used when two *M_Cons* operations are performed on a memory.

Theorems for the correctness of each top level transition are typified by the LDF transition theorem in Figure 3.11. The last 4 assumptions in the theorem match the applicable portion of *valid_program_constraint* for the LDF instruction branch. The major steps in proving this theorem include:

- deriving the value of the 9 bits of the 28 bit instruction code from the bottom assumption, for resolving with the less specific assumption of *LDF_state* and *LDF_Next*,

- rewriting the right side of the goal with the definition of *NEXT*,

- deriving the integer value of the 28 bit instruction code from the bottom assumption, and proving inequality of this value with all lesser values in order to reduce the right side of the goal to the *LDF_trans* branch,

- using *LDF_Next* to prove that *(TimeOf(is_major_state mpc)(t+1))* = *(TimeOf(is_major_state mpc)t)+26*

- rewriting the left side of the goal with *LDF_state* and the right side with the definition of *LDF_trans*

```
[ SYS_imp
; "clock_constraint SYS_Clocked"
; "reserved_words_constraint mpc memory"
; "well_formed_free_list mpc free memory"
; "mpc(TimeOf(is_major_state mpc)t) = #000101011"
; "is_cons(memory(TimeOf(is_major_state mpc)t)
          (c'(TimeOf(is_major_state mpc)t)))"
; "is_number(memory(TimeOf(is_major_state mpc)t)
              (car_bits(memory(TimeOf(is_major_state mpc)t)
                      (c'(TimeOf(is_major_state mpc)t)))))"
; "is_cons(memory(TimeOf(is_major_state mpc)t)
          (cdr_bits(memory(TimeOf(is_major_state mpc)t)
                  (c'(TimeOf(is_major_state mpc)t)))))"
; "atom_bits(memory(TimeOf(is_major_state mpc)t)
              (car_bits(memory(TimeOf(is_major_state mpc)t)
                      (c'(TimeOf(is_major_state mpc)t))))) =
  #00000000000000000000000000011" ]

⊢ s(TimeOf(is_major_state mpc)(SUC t)),
  e(TimeOf(is_major_state mpc)(SUC t)),
  c(TimeOf(is_major_state mpc)(SUC t)),
  d(TimeOf(is_major_state mpc)(SUC t)),
  free(TimeOf(is_major_state mpc)(SUC t)),
  mem_abs(memory(TimeOf(is_major_state mpc)(SUC t))),
  state_abs(mpc(TimeOf(is_major_state mpc)(SUC t))) =
  NEXT
  (s(TimeOf(is_major_state mpc)t),
   e(TimeOf(is_major_state mpc)t),
   c(TimeOf(is_major_state mpc)t),
   d(TimeOf(is_major_state mpc)t),
   free(TimeOf(is_major_state mpc)t),
   mem_abs(memory(TimeOf(is_major_state mpc)t))
```

Figure 3.11: The Correctness result for the LDF instruction

- using the abstract memory theorems of Figure 3.10 resolved with the assumptions about record types and the *free_list_constraint* to rewrite the right side of the goal, producing expressions for the next state in the same terminology as the left side.

- rewriting with the triplet selector function definitions (*cell_of, mem_of, free_of, mem_free_of,* etc.) used in the definition of LDF_trans.

Splitting the top goal (Figure3.5) on the state, and for *top_of_cycle* state, on the instruction codes permitted by the *valid_program_constraint*, reduces the problem to cases solved by similar theorems for each top level transition.

3.7   What Has Been Proved

If the goal of verifying hardware designs is to provide assurance that the designed (as opposed to fabricated) circuit will provide a defined functionality, the significance of the SECD verification effort must be judged in this overall context. The proof described has relates the register transfer level definition to the top level specification. A proof relating the low and register transfer levels is also being undertaken, but was not described herein. Taken together, these proofs only partially achieve the goal.

The top level specification lacks a definition for garbage collection. The proof has been constrained to cases where this is not required, but the issue of the bound in computation imposed by limited memory size has not been addressed. More generally, we have not given any assurance that the top level specification is consistent with the abstract SECD definition we started with.

The constraints (other than the *free_list_constraint*) represent reasonable limitations on the operation of the system. The *valid_program_constraint* will in the ideal case be guaranteed by a proven compilation process. As a start in this direction, a hand proof of the correctness of the compilation process of a high level language to SECD code has been completed. The rtl model (as also the low level model) is an abstraction of the lower level circuit. Although we believe that the model is correct, this belief is not supported (neither is it contradicted!) by this work.

The contributions made by this work include the use of a multi-bit data type in a large proof, the specification of a complex state relation using an abstract memory data type to represent complex S-expressions, and the achievement of a large proof result in the HOL system. Through the course of this project, we have learned that for a trained HOL user, designing proofs in the system, while at times exceedingly tedious, was not in itself that difficult. Managing the proof: the size of expressions, complexity of transitions, and number of cases, required great care. Most challenging, however, was deriving the logical representation of the machine and its specification, and formally defining the constraints.

Acknowledgements

We wish to acknowledge the insight into representing LISP data structures gained from conversations with Ian Mason ([19] and [20]). Inder Dhingra and Tom Melham gave us good advice and much help in discussions on the top level specification. We gratefully acknowledge the efforts of the SECD chip building team: Mark Brinsmead, Jeff Joyce (who began the development of the SECD in early 1985), Mary Keefe, Wallace Kroeker (Alberta Microelectronic Centre), Breen Liblong (Alberta Microelectronic Centre), Rick Schediwy, Konrad Slind, Glen Stone, Walter Vollmerhaus, Mark Williams, and Simon Williams.

The research described above was supported by Strategic, Operating, and Equipment Grants from the Natural Sciences and Engineering Research Council of Canada and The Canadian Microelectronics Corporation. The Strategic Grant was also supported by The Alberta Microelectronic Centre and LSI Canada Inc. The SECD verification effort was also supported by The Communication Research Establishment, Ottawa. We are also grateful to MOSIS (who built two versions of the chip) for their speedy, efficient and courteous service.

3.8   References

[1] François Anceau. *The Architecture of Microprocessors*. Addison-Wesley Publishing Company, 1986.

[2] G. Birtwistle, B. Graham, J. Joyce, S. Williams, M. Brinsmead, M. Keefe, W. Kroeker, B. Liblong, and W. Vollmerhaus. The SECD Machine on a Chip. In *Int. Conf. on CAD and CG*, Beijing, 1989.

[3] A. Church. A Formulation of the Simple Theory of Types. *Journal of Symbolic Logic*, 5:56–68, 1940.

[4] A. J. Cohn. A Proof of Correctness of the VIPER Microprocessors: The First Level. In G. Birtwistle and P. A. Subrahmanyam, editors,

VLSI Specification, Verification and Synthesis, pages 27–71, Norwell, Massachusetts, 1988. Kluwer.

[5] A. J. Cohn. A Proof of Correctness of the VIPER Microprocessors: The Second Level. In G. Birtwistle and P. A. Subrahmanyam, editors, *Trends in Hardware Verification and Automated Theorem Proving*, pages 1–91, New York, 1989. Springer Verlag.

[6] A. J. Cohn and M. J. C. Gordon. A Mechanized Proof of Correctness of a Simple Counter. Technical Report 94, Computing Laboratory, University of Cambridge, 1986.

[7] I. S. Dhingra. Formal Validation of an Integrated Circuit Design Style. In G. Birtwistle and P. A. Subrahmanyam, editors, *VLSI Specification, Verification and Synthesis*, pages 293–321, Norwell, Massachusetts, 1988. Kluwer.

[8] I. S. Dhingra. Formal Validation of an Integrated Circuit Design Style. PhD thesis, University of Cambridge Computer Laboratory, 1988.

[9] P. Gabriel. *Performance and Evaluation of LISP Systems*. MIT Press, Boston, 1985.

[10] M. J. C. Gordon. HOL: A Machine Oriented Formulation of Higher Order Logic. Technical Report 68, Computing Laboratory, University of Cambridge, 1985.

[11] B. Graham and G. Birtwistle. Formalising the Design of an SECD chip. In *Proceedings of the Cornell Workshop on Hardware Specification, Verification, and Synthesis: Mathematical Aspects*, New York, 1989. Springer Verlag.

[12] B. Graham, S. Williams, G. Birtwistle, J. Joyce, and B. Liblong. The Mossim Specification of the SECD DESIGN. Research report 89/341/03, Computer Science Department, University of Calgary, 1989.

[13] B. Graham, S. Williams, and G. Stone. Operating Specification for the SECD Chip. Research report 89/353/15, Computer Science Department, University of Calgary, 1989.

[14] Brian Graham. SECD: Design Issues. Research Report 89/369/31, Computer Science Department, University of Calgary, 1989.

[15] Brian Graham. Dealing with the Choice Operator in HOL88. Research Report 90/382/06, Computer Science Department, University of Calgary, 1990.

[16] P. Henderson. *Functional programming; applications and implementation.* Prentice Hall, London, 1980.

[17] J. Joyce. The SECD Machine, A Study in Advanced Architectures. Unpublished report of CPSC 603 course work at the University of Calgary.

[18] J. Joyce. Formal Verification and Implementation of a Microprocessor. In G. Birtwistle and P. A. Subrahmanyam, editors, *VLSI Specification, Verification and Synthesis*, pages 129–157, Norwell, Massachusetts, 1988. Kluwer.

[19] I. A. Mason. *The Semantics of Destructive Lisp.* Center for the Study of Languange and Information, Stanford, 1986.

[20] I. A. Mason. Verification of Programs that Destructively Manipulate Data. *Science of Computer Programming*, 10:177–210, 1988.

[21] T. F. Melham. Abstraction Mechanisms for Hardware Verification. In G. Birtwistle and P. A. Subrahmanyam, editors, *VLSI Specification, Verification and Synthesis*, pages 267–291, Norwell, Massachusetts, 1988. Kluwer.

[22] G. D. Plotkin. Call-by-name, call-by-value, and the lambda calculus. *Theoretical Computer Science*, 1(1):125–159, 1975.

[23] T. Simpson, G. Birtwistle, B. Graham, and M. J. Hermann. A Compiler for LispKit Targetted at Henderson's SECD Machine. Research Report 89/339/01, Computer Science Department, University of Calgary, 1989.

[24] T. Simpson, B. Graham, and G. Birtwistle. From Lispkit Code to SECD Chip: Some Steps on the Way to a Verified System. In *Proceedings of the Third Banff Higher Order Workshop*, 1989. submitted for publication.

[25] The HOL System. Tutorial. Technical Report, SRI International Cambridge Research Center, under contract to DSTO Australia, Cambridge, England, 1989.

[26] M. Williams. SECD Controller Board Implementation. Technical Report #89/359/21, Computer Science Department, University of Calgary, Calgary, 1989.

Formal Methods for VLSI Design
J. Staunstrup (Editor)
Elsevier Science Publishers B.V. (North-Holland)
© IFIP, 1990

Chapter 4

Formal Ruby

Lars Rossen[1]

4.1 Introduction

In other parts of these lecture notes, the relational language Ruby has been defined and we have investigated how to use a relational language for describing and reasoning about digital circuits.

In the following pages we shall see how we can give an alternative definition to the language that is suitable for implementing in a theorem prover. This new definition will give the same graphical interpretation to the Ruby expressions, but the algebraic definition will be different for some of the language constructs.

We will start by giving a slightly more complex definition of signals, followed by a definition of a core language, called Pure Ruby. These two sets of definition will be the basis for our axiomatic definition of Ruby in a theorem prover. After defining the Pure Ruby language, we will define the the rest of the Ruby language through Pure Ruby.

Finally some examples will be given to illustrate how the new definition can be used.

4.1.1 Notation

In the following sections we will use $\exists, \forall, \in, \wedge, \vee$ etc. with their usual meanings. When writing typed expressions we will use the notation $F: \alpha$ denoting

---

[1] Department of Computer Science, Technical University of Denmark, DK-2800 Lyngby, Denmark

that F has type α. For integers, natural numbers and booleans we use the type symbols $\mathbf{Z}, \mathbf{N}$ and $\mathbf{B}$ respectively. Following normal practice we will let relations be infix e.g. $a \, R \, b \ \equiv \ R(a,b) \ \equiv \ (a,b) \in R$ denotes; a is related to b through relation R. When defining relations in the Isabelle theorem prover we will use lambda abstractions, meaning that a relation between object of type α and β (a $\alpha \sim \beta$ relation) can be defined as a function (predicate) of type $\alpha \to \beta \to \mathbf{B}$. When we work with relations in this article we will use the infix notation. We will however define some relations using lambda abstraction. The notation of proofs is inspired by [3], and should be straight forward to read.

4.2   Signals and strong typing

In the normal definition of Ruby there is no information about the type of the ruby signal, but often there are some implicit restriction to the structure of the signals, e.g. when constructing a parallel composition, the signals the construction relates must be pairs. We will now introduce a strong typing on the Ruby expressions, and we start with a new definition of signals.

We will leave the basic concept of signals unchanged, that is signals will continue with being doubly infinite streams of data modelled as function from time (integers) to data:

$$\text{sig}(\alpha) \ = \ \mathbf{Z} \to \alpha$$

We will not give any restriction to the type of data in a signal, as with the original definition of Ruby we are interested in reasoning about the structure of the data. In this formulation of Ruby we drop the concept of tuples, and instead make a distinction between three kinds of data types: *primitive data*, *pairs of data* and *lists of data*. This gives us the following data type domain:

$$
\begin{aligned}
\text{datatype} \ = \ & \text{\textit{primitive datatype}} \\
| \ & (\text{\textit{datatype}} \times \text{\textit{datatype}}) \\
| \ & \text{list}_n (\text{\textit{datatype}})
\end{aligned}
$$

We are not concerned about what the *primitive datatype* can be, typically it will be integers, naturals, booleans, etc.

We shall see that pairs (a primitive form of tuples) and lists (with a length indicator) are a natural set of data constructors when defining Ruby formally.

As Ruby expressions relates signals, and not data, it is convenient to be able to reason about the structure of the data in the signals, and to do it without unpacking the data from the signals. This is exactly what signal tuples does to the underlying data tuples.

Therefore we introduce type constructors for signals, and corresponding to the three branches in the datatype domain (primitive, pairs, lists) we get three branches in the signaltype domains

Description	Type	Element constructor	Destructor
List with length n	$\text{list}_n(\alpha)$	nil $\text{cons}_n(a, b)$	$\text{hd}_n()$ $\text{tl}_n()$
Signal–list	$\text{List}_n(\alpha)$	Nil $\text{Cons}_n(a, b)$	$\text{Hd}_n()$ $\text{Tl}_n()$
Pair (binary tuples)	$(\alpha \times \beta)$	(a, b)	$\text{fst}()$ $\text{snd}()$
Signal–pair	$[\alpha \rtimes \beta]$	$[a, b]$	$\text{Fst}()$ $\text{Snd}()$

Figure 4.1: Signal types and constructors

$$
\begin{aligned}
\text{signaltype} \;=\; & \text{sig}(primitive\ datatype) \\
\mid\; & [signaltype \rtimes signaltype] \\
\mid\; & \text{List}_n(signaltype)
\end{aligned}
$$

The syntax for the type expressions, element construction, and destruction of the data and signal forms is shown in figure 4.1. It is possible to define the above types as a definitional extension to (standard) higher order logic. We will not show how to do that here.

Some abbreviations are needed in order to make it practical to write down signal expressions:

Lists:	$\{a_0, \cdots, a_n\}_{s(n)} \;=\; \text{cons}_n(a_0, \cdots \text{cons}_0(a_n, \text{nil}))$
Signal concatenation:	$a :_n b \;=\; \text{Cons}_n(a, b)$
Empty signal lists:	$\langle\rangle \;=\; \text{Nil}$
Signal–lists:	$\langle a_0, \cdots, a_n \rangle_{s(n)} \;=\; a_0 :_n \cdots a_n :_0 \langle\rangle$

Furthermore we need two more operators, one for appending lists (app), and one for appending signal lists (App):

List append:
$$l\, \text{app}_{0,m}\, \text{nil} \;=_{\text{def}}\; l$$
$$\text{cons}_n(h, l_1)\text{app}_{s(n),m} l_2 \;=_{\text{def}}\; \text{cons}_{n+m}(h, l_1\text{app}_{n,m}l_2)$$

Signal append:
$$l\, \text{App}_{0,m}\, \text{Nil} \;=_{\text{def}}\; l$$
$$\text{Cons}_n(h, l_1)\text{App}_{s(n),m} l_2 \;=_{\text{def}}\; \text{Cons}_{n+m}(h, l_1\text{App}_{n,m}l_2)$$

4.3 Pure Ruby

The next step is to define what a Ruby relation is. In general it is a relation between signals ($\text{sig}(\alpha) \sim \text{sig}(\alpha)$), but there are a lot of mathematically well defined relations over signals that we don't accept as Ruby relations. Therefore we want to make a precise statement of what a Ruby relation is.

To make an axiomatisation of Ruby easy, we will select a small set of primitive Ruby constructs, define them in the logic, and then define the rest of the Ruby language in term of these primitives. Later we will see how a small set of primitives helps us prove general things about Ruby expressions.

In this axiomatisation of Ruby we will use a syntactic domain for Ruby relations which permits four basic constructs. The language defined by this syntax will be called Pure Ruby. The domain is defined by:

$$
\begin{aligned}
\text{ruby} = \ & \text{spreads}(f) \\
| \ & \mathcal{D} \\
| \ & ruby;ruby \\
| \ & [ruby, ruby]
\end{aligned}
$$

These primitives correspond to *combinational circuits, delay element, serial composition* and *parallel composition* (notice that we have redefined spread to be a (higher order) function from relation to stream relation).

As previously noted we will define the rest of the Ruby language in terms of the above Pure Ruby, but there are a few primitives in Ruby described in [9] that can not be defined through Pure Ruby. Examples of primitives that can't be defined in terms of Pure Ruby are bundle and skew. These primitives are used when reasoning about circuits with more than one time base. We have deliberately chosen not to include these in Pure Ruby as they don't conform to some nice algebraic properties. It is important to note that though these primitives are not part of Pure Ruby, it is perfectly all right to include them in a specification, and in our proof system, as long as we make a distinction between specification with and without these extra primitives.

A good way to make this distinction is to introduce a type that corresponds to our Pure Ruby domain.

4.3.1   Pure Ruby type

Before we discuss how to implement a Pure Ruby type, we will give the definitions of the four Pure Ruby constructs. In these definitions information on the type of signal the construct relates are given, and it is this type information that ensures that the relations are well defined for all arguments (of the given type).

The definitions of the constructors of Pure Ruby are:

Axiom

Spread $\vdash$ a: $\text{sig}(\alpha)$ $\text{spreads}(f\colon \alpha \sim \beta)$ b: $\text{sig}(\beta)$ $=_{\text{def}}$ $\forall t \cdot a(t) \, f \, b(t)$

Delay $\vdash$ a: $\text{sig}(\alpha)$ $\mathcal{D}$ b: $\text{sig}(\beta)$ $\qquad\quad =_{\text{def}}$ $\forall t \cdot a(t-1) = b(t)$

Ser. $\vdash$ a: $\text{sig}(\alpha)$ $F;G$ b: $\text{sig}(\beta)$ $\qquad =_{\text{def}}$ $\exists c$: $\text{sig}(\gamma) \cdot a \, F \, c \, \wedge \, c \, G \, b$

Par. $\vdash$ a: $[\alpha_1 \bowtie \alpha_2]$ $[F, G]$ b: $[\beta_1 \bowtie \beta_2]$ $=_{\text{def}}$

$\qquad\qquad\qquad\qquad\qquad\qquad \text{Fst}(a) \, F \, \text{Fst}(b) \, \wedge \, \text{Snd}(a) \, G \, \text{Snd}(b)$

The above definitions classify the four constructs as being signal relations, but as mentioned before we want to introduce a type that express that this is the only way to make a Pure Ruby (signal) relation.

If we use the syntactic notation $\alpha \overset{Pure}{\sim} \beta$ to denote "Pure Ruby relation", then we have the following domain law:

$$\alpha \overset{Pure}{\sim} \beta \subset \text{sig}(\alpha) \sim \text{sig}(\beta)$$

This law simply states that the domain of Pure (Ruby) relations is a subset of the domain of signal relations in general.

To give the $\alpha \overset{Pure}{\sim} \beta$ symbol a constructive definition we state the following type axioms about Pure Ruby.

Axiom

Spread-type	$f\colon \alpha \sim \beta$	$\vdash\ \text{spreads}(f)\colon \alpha \overset{Pure}{\sim} \beta$
Delay-type		$\vdash\ \mathcal{D}\colon \alpha \overset{Pure}{\sim} \alpha$
Ser.-type	$F\colon \alpha \overset{Pure}{\sim} \beta,\ G\colon \beta \overset{Pure}{\sim} \gamma$	$\vdash\ F;G\colon \alpha \overset{Pure}{\sim} \gamma$
Par.-type	$F\colon \alpha_1 \overset{Pure}{\sim} \beta_1,\ G\colon \alpha_2 \overset{Pure}{\sim} \beta_2$	$\vdash\ F;G\colon (\alpha_1 \times \alpha_2) \overset{Pure}{\sim} (\beta_1 \times \beta_2)$

From this Ruby type definition we can see that **Spread** and **Delay** ($\mathcal{D}$) are the primitive relations when constructing Ruby expressions and the other two construct Ruby expressions from other Ruby expressions.

This observation leads to the definition of a Ruby induction theorem.

Ruby Induction

$\forall f\colon \alpha \sim \beta \cdot \ \text{P}(\text{spreads}(f)) \ \wedge$

$\text{P}(\mathcal{D}) \ \wedge$

$\forall F\colon \alpha_1 \overset{Pure}{\sim} \beta_1, G\colon \alpha_2 \overset{Pure}{\sim} \beta_2 \cdot \ \text{P}(F) \wedge \text{P}(G) \ \Rightarrow \ \text{P}(F;G) \ \wedge$

$\forall F\colon \alpha \overset{Pure}{\sim} \beta, G\colon \beta \overset{Pure}{\sim} \gamma \cdot \ \text{P}(F) \wedge \text{P}(G) \ \Rightarrow \ \text{P}([F,G])$

$\rule{8cm}{0.4pt}$

$\forall R\colon \alpha \overset{Pure}{\sim} \beta \cdot \ \text{P}(R)$

4.3.2 Ruby–extension

The next step in the process of making a Ruby system is to define the rest of the standard Ruby combining forms. It is not within the scope of these notes to define all the forms that the Ruby language consist of, but below we will define a set of wiring relations, and combining forms that covers a reasonable subset of Ruby.

The first is the definition of relational inverse. The normal definition is:

Theorem: Inverse $a\ F^{-1}\ b\ =\ b\ F\ a$

If we use that definition we could not use the Ruby type axioms on expressions involving inverse without extending Pure Ruby with that form. Instead we

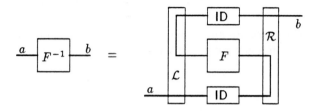

Figure 4.2: Inverse definition

define inverse through Pure Ruby. First we define three spreads; $\mathcal{R}$, $\mathcal{L}$ and ID:

Definitions.

$\mathcal{L}$-def. $\quad \mathcal{L} =_{\text{def}} \text{ spreads}(\lambda ab \cdot \exists c \cdot b = (a, (c, c)))$

$\mathcal{R}$-def. $\quad \mathcal{R} =_{\text{def}} \text{ spreads}(\lambda ab \cdot \exists c \cdot a = (c, (c, b)))$

ID-def. $\quad \text{ID} =_{\text{def}} \text{ spreads}(\lambda ab \cdot a = b)$

If we look at figure 4.2 its seem reasonable to define Inverse as:

Definition: Inverse $F^{-1} =_{\text{def}} \mathcal{L}; [\text{ID}, [F, \text{ID}]]; \mathcal{R}$

To complete this definition one has to prove that the normal meaning of inverse is true from the definition. Furthermore it is convenient to prove a theorem stating that the inverse of a Pure Ruby relation is also a Pure Ruby relation.

This was gentle introduction to the idea of defining Ruby in terms of Pure Ruby. We now state a set of definitions for wire relations and combining forms.

- **Pair1 & 2** is used for hiding signals in a pair:

$$\pi_1 =_{\text{def}} \text{ spreads}(\lambda ab \cdot a = \text{fst } b)$$
$$\pi_2 =_{\text{def}} \text{ spreads}(\lambda ab \cdot a = \text{snd } b)$$

- **Join** is the wiring primitive that joins 4 wires together, it is define through a spread:

$$join: (\alpha \times \alpha) \overset{Pure}{\sim} (\alpha \times \alpha) =_{\text{def}} \text{ spreads}(\lambda ab \cdot \text{fst } a = \text{snd } a = \text{fst } b = \text{snd } b)$$

- **Apply left** relates a pair consisting of an element and a list, with the list consisting of the element appended to the left side of the list:

$apl_n \colon (\alpha \times \mathsf{list}_n(\alpha)) \overset{Pure}{\sim} \mathsf{list}_{s(n)}(\alpha) =_{\mathrm{def}}$
$$\mathsf{spreads}(\lambda ab \cdot \ b = (\{\mathsf{fst}\,a\}_1)\mathsf{app}_{1,n}(\mathsf{snd}\,a))$$

- **Apply right** relates a pair consisting of a list and an element, with the list consisting of the element appended to the right side of the list:

$apl_n \colon (\alpha \times \mathsf{list}_n(\alpha)) \overset{Pure}{\sim} \mathsf{list}_{s(n)}(\alpha) =_{\mathrm{def}}$
$$\mathsf{spreads}(\lambda ab \cdot \ b = (\mathsf{fst}\,a)\,\mathsf{app}_{n,1}\{\mathsf{snd}\,a\}_1)$$

- **Reorg** makes the following regrouping of pairs: $[[a,b],c]\mathsf{reorg}[a,[b,c]]$ but again the definition is done through a spread:

$\mathsf{reorg} \colon ((\alpha \times \beta) \times \gamma) \overset{Pure}{\sim} (\alpha \times (\beta \times \gamma)) =_{\mathrm{def}}$
$$\mathsf{spreads}(\lambda ab \cdot \ \mathsf{fst}(\mathsf{fst}(a)) = \mathsf{fst}(b) \ \wedge$$
$$\mathsf{snd}(\mathsf{fst}(a)) = \mathsf{fst}(\mathsf{snd}(b)) \ \wedge$$
$$\mathsf{snd}(a) = \mathsf{snd}(\mathsf{snd}(b)))$$

- **Swap** is a wiring primitive that swaps two wires without joining them:

$\mathsf{swap} \colon (\alpha \times \beta) \overset{Pure}{\sim} (\beta \times \alpha)$
$$=_{\mathrm{def}} \mathsf{spreads}(\lambda ab \cdot \ \mathsf{fst}(a) = \mathsf{snd}(b) \ \wedge \ \mathsf{snd}(a) = \mathsf{fst}(b))$$

- **NIL** is a primitive that relates two empty lists of any type. It is a primitive that is convenient to have when defining some of the combining forms.

$\mathsf{NIL} \colon \mathsf{list}_0(\alpha) \overset{Pure}{\sim} \mathsf{list}_0(\alpha) =_{\mathrm{def}} \mathsf{spreads}(\lambda ab \cdot \ (a = b = \mathsf{nil}))$

- **Conjugate** continues to have its traditional definition:

$$R \setminus S =_{\mathrm{def}} S;R;S^{-1}$$

- **Map** is a generic combining form that expands according to the length of the signal it relates, we shall therefore define it in terms of recursion:

$\mathsf{map}_0(F) \quad =_{\mathrm{def}} \mathsf{NIL}$
$\mathsf{map}_{s(n)}(F) \quad =_{\mathrm{def}} [F, \mathsf{map}_n(F)] \setminus apl_n^{-1}$

- **Beside** is defined through the use of **Reorg, Parallel composition** and **ID**:

$$P \leftrightarrow Q =_{\mathrm{def}} \mathsf{reorg}^{-1};[P,\mathsf{ID}];\mathsf{reorg};[\mathsf{ID},Q];\mathsf{reorg}^{-1}$$

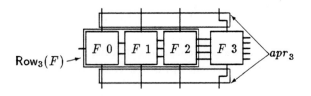

Figure 4.3: row definition

- **Below** is defined through the use of **Inverse** and **Beside**:

$$P \updownarrow Q \ =_{\text{def}} \ (P^{-1} \leftrightarrow Q^{-1})^{-1}$$

- **Row** is defined recursivly as **map**, but it is a bit more complicated. The trick is to understand that F is a function from a position number to a Ruby relation. This way it is possible to define more complicated forms than if each F in the row had to be the same.

$$\begin{aligned} \mathsf{Row}_0(F) \quad &=_{\text{def}} \ [\mathsf{ID}, \mathsf{NIL}]; \mathsf{swap} \\ \mathsf{Row}_{s(n)}(F) \quad &=_{\text{def}} \ \mathsf{Snd}(apr_n^{-1}); (\mathsf{Row}_n(F) \leftrightarrow (F \ n)); \mathsf{Fst}(apr_n) \end{aligned}$$

An example of the step case can be seen in figure 4.3.

Most of the above definitions are rather obscure when seen for the first time. One has to remember though that they are only definitions and that the usual properties can be derived from these definitions. In fact one should forget these definitions as soon as theorems stating the normal meaning has been proven.

We will show how to do this in the case of the combining form **map**. We will normally need 2 theorems, one that states that $\mathsf{map}_n(F)$ is a Pure Ruby relation whenever F is, and one that states the meaning of **Map** in terms of a decomposition of the signals it relates. The two theorems called **map-type** and **map** are:

Theorem

Map-type: $F\colon \alpha \overset{Pure}{\sim} \beta \ \Rightarrow \ \mathsf{map}_n(F)\colon \mathsf{list}_n(\alpha) \overset{Pure}{\sim} \mathsf{list}_n(\beta)$

Map: $\mathsf{Nil} \ \mathsf{map}_0(F) \ \mathsf{Nil}$

$$a{:}_n b \, \mathsf{map}_{s(n)}(F) c{:}_n d \ = \ a F c \ \wedge \ b \, \mathsf{map}_n(F) d$$

A proof for Map-type and the base case of Map is elementary. Before we show a simple proof for the inductive case we state the theorem that captures the usual interpretation of *apl*.

Theorem: Apl $\vdash [a,b]\,apl_n\,a:_n b$

The proof of **map** is:

Proof

From $n \in \mathbb{N}$

1	$a:_n b\,\mathsf{map}_{\mathsf{S}(n)}(F)\,c:_n d \;=\; a:_n b\,\mathsf{map}_{\mathsf{S}(n)}(F)\,c:_n d$	=-Refl.
2	$a:_n b\,\mathsf{map}_{\mathsf{S}(n)}(F)\,c:_n d \;=$	map-def.
	$a:_n b\,apl_n^{-1}\setminus[F,\mathsf{map}_n(F)]\,c:_n d$	
3	$a:_n b\,\mathsf{map}_{\mathsf{S}(n)}(F)\,c:_n d \;=$	cunj-def.
	$a:_n b\,apl_n^{-1};[F,\mathsf{map}_n(F)];apl_n\,c:_n d$	
4	$a:_n b\,\mathsf{map}_{\mathsf{S}(n)}(F)\,c:_n d \;=\; \exists e_1 e_2 \cdot$	Inv.,;-def
	$(e_1\,apl_n\,a:_n b)\;\wedge\;(e_1[F,\mathsf{map}_n(F)]e_2)\;\wedge\;(e_2\,apl_n\,c:_n d)$	
5	$a:_n b\,\mathsf{map}_{\mathsf{S}(n)}(F)\,c:_n d \;=\; [a,b][F,\mathsf{map}_n(F)][c,d]$	apl_n
Infer	$a:_n b\,\mathsf{map}_{\mathsf{S}(n)}(F)\,c:_n d \;=\; aFc\;\wedge\;b\,\mathsf{map}_n(F)\,d$	Par-def.

4.4 Implementation

As mentioned previously the goal of this treatment of Ruby is to implement the Ruby algebra in a system supporting formal reasoning. We have done so in the Isabelle theorem prover. The Isabelle theorem prover is a very flexible system, that support a wide variety of logics. The system is described in [4][6][7]. The Isabelle system defines a simple Meta-Logic. Object logics are then defined in this (meta-) logic.

One of the object logics distributed with the system is a Higher Order Logic (HOL) and the Ruby system is defined as an extension to this logic. This logic is a many sorted logic similar to the logic in the HOL88 system [1][2]. A main difference is that in HOL88 the typing of expressions is decideable and the typing is done automatically by the term parser. In Isabelle-HOL the type information is part of the algebra and one has to reason about the type of terms. This concept is related to the idea of "Propositions as type" and gives Isabelle/HOL a more advanced type system. This makes it possible directly to implement the length dependent type $\mathsf{list}_n(\alpha)$. It also makes the proof obligation bigger, as one has to prove theorems about types. Refer to [5] for a discussion of Isabelle types. The actual implementation is described in [8], where it is possible to find some examples on the use of the system.

As we have mentioned, the axiomatisation as it has been described here is tailored for an implementation in a theorem prover. Below we will go through the main differences between this axiomatisation and the original definition.

- The first thing we did was to introduce pairs and lists instead of tuples. This was done to make it possible to fit the type of signals into a polymorphic type system. The general tuple type would in a higher order logic introduce inconsistency.

- When we defined the list type we did it as a parametrised type over the length of the list. That way it was possible to give a precise type to the combining forms that relates two lists of equal length, e.g. **map** is only well-defined for signals lists of equal length.

- Finally we selected a small subset of the Ruby language called Pure Ruby, and defined the rest of the Ruby language through this basis. We introduced the Pure Ruby type to encapsulate the definition, thereby making it possible to state a Ruby induction theorem, used for proving general things about Ruby.

Below is a theorem that states a general property of all Pure Ruby expressions, and it is proved through the use of the Ruby induction theorem.

4.5   Example

The *retiming* theorem is a good example of a theorem that is essential in the synthesis process of systolic circuits [9]. The retiming theorem states that the relational behaviour of a circuit is preserved when surrounding it with delay elements (fig. 4.4). There are some kinds of circuits that don't have this property, but one of the nice features of Pure Ruby is that it always have the retiming property:

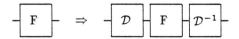

Figure 4.4: **Retime** definition

Definition Retime: $\vdash$ Retime F $=_{\text{def}}$ $(\text{P}(F) \Rightarrow \text{P}(\mathcal{D};F;\mathcal{D}^{-1}))$
Theorem Retime: $\vdash$ $\forall$ F: $\alpha \overset{Pure}{\sim} \beta$ $\cdot$ Retime F

To prove this theorem we prove the corresponding theorem for each of the four forms of Pure Ruby and then we use the Ruby–Induction rule. The four theorems are:

Theorem
Retime Spread $\vdash$ $\forall$ f $\cdot$ Retime spreads(f)
Retime Delay $\vdash$ Retime $\mathcal{D}$
Retime Comp $\vdash$ (Retime F $\wedge$ Retime G) $\Rightarrow$ Retime ($F;G$)
Retime Par $\vdash$ (Retime F $\wedge$ Retime G) $\Rightarrow$ Retime $[F, G]$

By defining all Ruby constructs in terms of Pure Ruby it is possible to prove general rules in an easy way. To use the above retiming rule the only thing to do before rewriting a subexpression with a retimed form is to prove that the expression has the type $\alpha \overset{Pure}{\sim} \beta$.

A useful lemma when proving the above 4 theorems is:

Lemma: Delay-ID $P(\mathcal{D};\mathcal{D}^{-1}) \Rightarrow P(\mathsf{ID})$

A proof for this is:

Proof

From $P(\mathcal{D};\mathcal{D}^{-1})$

1	$P(\lambda ab \cdot \exists c \cdot (b \; \mathcal{D} \; c) \wedge (c \; \mathcal{D}^{-1} \; b))$	
2	$P(\lambda ab \cdot \exists c \cdot (\forall t \cdot a(t-1) = c(t)) \wedge (\forall t \cdot b(t-1) = c(t)))$	Inv,$\mathcal{D}$-def.
3	$P(\lambda ab \cdot \exists c \cdot ((\lambda t \cdot a(t-1)) = c) \wedge ((\lambda t \cdot b(t-1)) = c))$	$\forall$-elim.
4	$P(\lambda ab \cdot (\lambda t \cdot a(t-1)) = (\lambda t \cdot b(t-1)))$	$\exists$-elim.
5	$P(\lambda ab \cdot a = b)$	rew.
Infer	$P(\mathsf{ID})$	ID-theorem.

We will end this example with the proof of the **Retime Comp** theorem:

Proof

From Retime $F \wedge$ Retime G

1	$P'(F) \Rightarrow P'(\mathcal{D};F;\mathcal{D}^{-1})$	Retime-Def.
2	$P(F;G) \Rightarrow P(\mathcal{D};F;\mathcal{D}^{-1};G)$	Inst-P'
3	$P''(G) \Rightarrow P''(\mathcal{D};G;\mathcal{D}^{-1})$	Retime-Def.
4	$P(\mathcal{D};F;\mathcal{D}^{-1};G) \Rightarrow P(\mathcal{D};F;\mathcal{D}^{-1};\mathcal{D};G;\mathcal{D}^{-1})$	Inst-P''
5	$P(F;G) \Rightarrow P((\mathcal{D};F;\mathcal{D}^{-1});(\mathcal{D};G;\mathcal{D}^{-1}))$	$\Rightarrow$Trans 2,4
6	$P(F;G) \Rightarrow P(\mathcal{D};F;(\mathcal{D}^{-1};\mathcal{D});G;\mathcal{D}^{-1})$	Comp-ass.
7	$P(F;G) \Rightarrow P(\mathcal{D};F;\mathsf{ID};G;\mathcal{D}^{-1})$	delay-ID
8	$P(F;G) \Rightarrow P(\mathcal{D};F;G;\mathcal{D}^{-1})$	ID-elim.
Infer	Retime $(F;G)$	Retime-Def.

4.6 Summary

In the previous section we have seen how to reformulate the definition of Ruby in a way suitable for implementing in a theorem prover.

First we defined a more differentiated signal type, consisting of both pairs and lists of signals.

With these types at hand we selected a small number of Ruby constructs as a Pure Ruby system. We formally define the universe of Pure Ruby constructs by introducing a Pure Ruby type.

Finally we went on to define the rest of the Ruby universe through the Pure Ruby construct. By defining Ruby through a few primitives it has become possible to prove general properties of the Ruby language, by proving that the property holds for the Pure Ruby constructs.

4.7   References

[1] Mike Gordon. Hol a machine oriented formulation of higher order logic. Technical Report 68, University of Cambridge, 1986.

[2] Mike Gordon. Hol a proof generating system for higher-order logic. *VLSI Specification, Verification and Synthesis*, 1987.

[3] Cliff B. Jones. *Systematic Software Development Using VDM*. Prentice/Hall International, 1986.

[4] Lawrence C. Paulson. Natural deduction as higher-order resulution. *The Journal of Logic Programming*, 3, 1986.

[5] Lawrence C. Paulson. A formulation of the simple theory of types. Technical report, Computer Laboratory, University of Cambridge, 1989.

[6] Lawrence C. Paulson. The foundation of a generic theorem prover. *Journal of Automated Reasoning*, 5, 1989.

[7] Lawrence C. Paulson and Tobias Nipkow. *Isabelle Tutorial and User's Manual*, 1990.

[8] Lars Rossen. Isabelle-Ruby, tutorial and reference guide. Technical report, Dept. of Computer science Technical University of Denmark, 1990. In preperation.

[9] Mary Sheeran. Retiming and slowdown in ruby. *Proceedings, Glasgow workshop on the Fusion of Hardware Design and Verification*, 1988.

Formal Methods for VLSI Design
J. Staunstrup (Editor)
Elsevier Science Publishers B.V. (North-Holland)
© IFIP, 1990

Chapter 5

Formal System Design

M.P. Fourman[1]

5.1 Introduction

Simulation is necessary, but inadequate. System level design requires treatment of behaviour at many levels of abstraction. Design and verification must be integrated.

Engineering may be described as the application of scientific laws to the design and implementation of useful systems. Laws allow the engineer to proceed; he moves from a specification to an implementation satisfying the specified requirements. Laws allow him to reason about (models of) the system. The better the models, and the better the calculi for reasoning about them, the less he needs to resort to trial and error — and the better the software tools we can build to support him. In this chapter, we introduce the use of formal logic for the *behavioural* specification and design of digital systems.

5.1.1 Modelling

To design an electric heater, we model the heater element as a pure resistance. Ohm's law allows us to calculate the current which will flow for a given voltage and resistance. To implement a state machine using a PLA (programmed logic array) we model the PLA as a Boolean function. We use Boolean algebra to arrive at a suitable function and to optimise the PLA. Most such models are so ingrained that we are liable to forget that they are only models.

---

[1] Addr.: Dept. of Computer Science, JCMB, King's Buildings, Edinburgh University, Edinburgh EH9 3JZ, Scotland, UK. e-mail: mikef@lfcs.ed.ac.uk

The design engineer must consider many aspects of a system. At the electrical level, he deals with voltages, currents and power. These are modelled as continuous, real-valued functions of time; we use Maxwell's electromagnetic theory and the infinitesimal calculus of Leibnitz and Newton to reason about them. At the physical level, he deals with such properties as size and power consumption. We are concerned with the *behaviour* of digital systems at a level which abstracts away these electrical and physical aspects.

The behaviour of a digital system can be complex. To manage this complexity, we must consider many levels of behavioural *abstraction*. Many specialised calculi have been introduced for describing and reasoning about system behaviour at particular levels — for example, ISP [3], CCS [29], CSP [17], Interval Temporal Logic [31]. Higher-order, predicate logic is a general-purpose formalism that allows us to describe and relate both behaviour and stucture at many different levels [13]. This chapter will show how we can use it to support behavioural design at higher levels of abstraction.

5.1.2   Specification

Specifications are essential contractual and methodological tools. The primary use of a behavioural model is to *specify* the required behaviour. This specification is central to the work of an engineer; his basic task is to design a system that meets the specification. Because the specification provides the interface between customer and designer, it must be understood by both.

> The Eskimo, have many words for distinct types of snow. In English, we do not have words for these distinctions. To translate an Eskimo treatise on building with snow, we should have to first explain these concepts. I imagine that, for parts of the body, the Eskimo have much the same vocabulary as we. It would be much easier to translate an Eskimo treatise on splinting fractured limbs.

The customer and designer should have a common language, using a vocabulary of shared concepts to give, and understand, the specification. Sharing just the vocabulary (syntax) is not enough; they must also agree on meaning (semantics). As well as agreeing on primitive concepts, they must use the same means of constructing statements about the concepts (logic) and agree on sound methods for reasoning about them (proof).

We are interested in specifications; writing them and reasoning about them. The following points should be born in mind when writing a specification:

- Specification is not implementation. A specification should constrain the implementation as little as possible. There may be a component available on the market (or in your library) that will do the job. If you over-specify, you could, nevertheless, end up with an expensive custom part.

- Specification should use a language common to specifier and implementer. Decide which concepts the specifier and implementer have in common. Use as few primitives as possible. This will make it easier to check that the specification is understood in the same way by specifier and implementer (and will make this more likely). It will also make it easier to formalize the specification later on. Generally, the implementer will have to *design* concrete realizations for the abstract data of the specification.

- Express the requirements concisely and precisely. It should be possible to satisfy the requirements. It should be possible to decide objectively whether the requirements are satisfied by a given design. It should be possible to understand the requirements.

- Make sure *you* understand what you want. (It helps to do this first!)

5.1.3 Abstraction

Specifications have another, equally important, rôle: they express abstraction. The specification of a component should allow us to use that component without worrying about (or even knowing about) its inner workings. This *behavioural abstraction* is crucial to the design of large systems; it provides the basis for managing complexity using hierarchical decomposition. Specifications are also necessary for scientific project management; once the specification of a component has been fixed, its users and implementers can work in parallel. Well-specified components also form the basis for reusable libraries.

Product specification and design documentation are often confused. A *specification* specifies a behaviour which may be implemented in different ways. Any user relying on the specification should be able to substitute one implementation for another without affecting the correctness of the overall system. Design documentation describes an implementation, it is necessary for maintenance and documents features incidental to the particular implementation. Using design documentation as a basis for incorporating a component in a system is a mistake — if you rely on an incidental feature you have no freedom to improve your system design by later substituting a different implementation of the component specification, unless, by chance, it shares the 'feature'.

We also use *interface specifications* to express the *data abstraction* and *temporal abstraction* which relate the concrete data and time of an implementation to the abstractions of the design specification.

5.1.4   Validation

To make the contract between engineer and customer effective, it must be possible to check whether or not the design produced satisfies the requirements specified. The *acceptance criteria* employed give contractual meaning to the specification. How do we check that the design meets the specification? It may be possible to test a sample of the final product to see if it satisfies the specification — we might use such a method to test the design of a fire-door. This is a simple acceptance criterion based on product sampling. It is an approach we would be unlikely to use in validating the design of a road bridge. We might, however, build a model bridge, which should be cheaper than building a full-size bridge, and subject it to tests. The customer might agree to accept the design provided the performance of the model is satisfactory. This is a simple example of acceptance based on simulation. Its success relies on understanding the relationship between the model and reality.

Another approach, common in mature engineering disciplines, is to use abstract models which enable us to predict the behaviour of a design. The existence of abstract models distinguishes engineering from craft. Using these models, it is not necessary to rely on a crude 'suck it and see' approach. The designer can formulate a solution and the customer can, in principle, check the calculations which show that the requirement is satisfied. We call this *design verification.* The success of this approach relies on understanding the relationship between the model and reality, *and* on the existence of a tractable calculus for reasoning about the model.

In digital systems engineering, 'breadboarding' (building samples for design validation) is no longer feasible: the cost of producing a set of masks for a chip design is high and any modifications require a complete reiteration. Because of this it is customary to simulate software models, which are cheaper to produce and, more important, easy to modify. Unfortunately, simulation models usually run much more slowly than the system they simulate.

The models we use are always abstractions of reality. A circuit-level simulation is based on models at the electrical level. A switch-level model idealizes transistors as switches. A gate-level model idealizes collections of transistors as Boolean functions. A functional model might idealize a complete block as a multiplier (or whatever). We gain power by simulating at this higher, functional level. Before we can have confidence in such a high-level simulation, we have to satisfy ourselves that these abstractions are valid.

Run-times for simulation soon become excessive, as complexity rises exponentially with the number of inputs and internal states. In any case, it is usually impossible to establish by simulation or test that a design meets its specification: There are too many possible combinations of input sequences. A 32-bit multiplier has 2^{64} possible inputs. There are approximately 2^{45}

nanoseconds in a year. At best we can simulate a small proportion of the possible inputs. Partial simulations can only give partial validations. In these circumstances, we are forced to reason about an abstract model.

One powerful approach is *symbolic simulation* [4, 5]; we can simulate with boolean *expressions* (instead of boolean values), and use algebraic manipulations (typically, rewriting to a normal form) to compare the output expressions with those specified. This is the best approach available for validation of combinational logic. Symbolic simulation automates simple equational reasoning (typically about boolean values). For higher levels of design, algebraic comparisons become intractable. More flexible formalisms are needed to express specifications, and multi-level reasoning is needed to relate specification to implementation *via* abstractions. Reasoning at various levels of abstraction allows the designer great freedom in investigating the relationship between specification and possible implementations. Formal predicate logic with inductively-defined and higher-order types provides the necessary expressive power, and formal proof allows us to manipulate behaviours.

The contractual obligation of the designer is currently based on test suites which are necessarily inexhaustive, and are themselves hard to validate. Formal verification of the logical correctness of a design could provide a more satisfactory basis for the *design* contract[2]. Formal verification can guarantee logically correct design. This provides added value in safety-critical applications. There will also be benefits in reducing time to market; to be 'right first time' is of paramount importance.

5.2   Modelling Digital Behaviour

Every simulator embodies an operational model (usually state-based, often implicit, and frequently ill-understood) of behaviour. Reasoning about such models is like reasoning about imperative programs. At an appropriate level of abstraction, we can use mathematical models of behaviour, better-adapted to reason. In this section, we present such a model. To emphasize the familiarity of our approach, we look briefly at an unremarkable example of modelling from a different area.

Example

To model the behaviour of a system connecting a power source to a load, we plot current against voltage, as shown in Figure 5.1. The curves represent the combinations of voltage and current provided by the source, and consumed by the load. When we connect these subsystems, they both impose constraints

---

[2] Test suites will remain essential for production test, and simulation for validating specifications

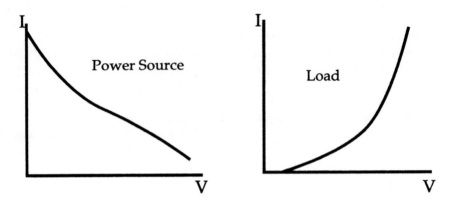

Figure 5.1:

relating voltage and current.   The resulting possibilities are given by inter-

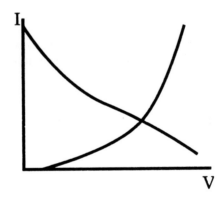

Figure 5.2: Finding the combined behaviour.

secting the two graphs (Figure 5.2). Note that the source and the load are
treated equivalently. The formalism does not capture the causal asymmetry
between them.

Of course, graphical methods only serve to complement analytic methods.
Given equations for the two curves, we can use algebra to find their intersection. Graphical methods make the interaction of the two subsystems easy
to visualise. Analytic methods are more generally useful. From an abstract
point of view, the behaviour of each subsystem is modelled as a set of pairs
(the points on the graph of the function); to model the composite system, we
intersect these sets. Our model of digital behaviour will be formally similar.

5.2.1   Functional Models

Some circuits may be modelled straight-forwardly as functions.   Our first model of circuit behaviour is based on Boolean functions and equations. We model a complex gate G, with input ports $\mathbf{I}$ and output ports $\mathbf{O}$ by an equation

$$\mathbf{O} = f(\mathbf{I}),$$

where f is a vector of Boolean functions, and $\mathbf{I}$ and $\mathbf{O}$ are just vectors of Boolean variables.

This vector of equations may be regarded as the *conjunction* of its component equations.   The basic method of combining circuits is to add extra conjuncts.   To reason about behaviour we use *Boolean algebra* to manipulate the equations.   This certainly *seems* simple enough — but if we consider the circuit given by

$$c = and(a, b) \;\&\; c = or(a, b)$$

for the input values $a = hi$, $b = lo$, we obtain $c = hi \;\&\; c = lo$, which is nonsense and reveals a limitation of our model.   What this circuit actually *does* will depend on the technology used to implement the gates.   When designing at the gate level we usually impose the *design rule*, 'do not connect outputs together,' which prohibits this example.   We must also impose the rule, 'do not create feedback loops,' (otherwise we could connect the two ports of an inverter to get $x = not\ x$ in the model).   Similar functional models are also used (particularly to analyse 'data-flow' architectures) at higher levels of abstraction; similar design rules apply.

Sometimes,*when we wish to take advantage of behaviour at a lower level of abstraction, we break the rules*, on purpose.   For example, we may know, from consideration of driving ratios, that connecting outputs has the effect of a *wired* **and**, or that, because of gate delays, we can make a flip-flop with cross-coupled nand gates.   If we obey the rules then the functional model and equational reasoning will not lead us astray; when we want to break them, we have to pay a price, and use more detailed models.

Using conjunctions of equations to model behaviour does not allow us enough flexibility when we wish to *abstract* the behaviour of composite components.   Consider Figure 5.3.   We connect an **and** and an **or** gate; the behaviour is

$$d = a\ and\ b \;\&\; e = d\ or\ c.$$

If we want to abstract the structure and view this component as a black box (Figure 5.4),   we would like to arrive at the behaviour

$$e = (a\ and\ b)\ or\ c.$$

This is achieved by function composition — but we have also made d invisible. We want a formalism which allows us to express such *hiding* of internal wires, and other ways of combining circuits.

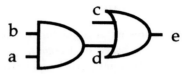

Figure 5.3: Two gates.

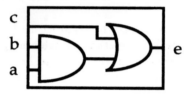

Figure 5.4: A black box.

5.2.2 The Relational Model

In our electrical example, each sub-system imposed constraints on the other. To determine the behaviour of the whole we had to conjoin the constraints imposed by the subsystems. We now approach the behaviour of digital systems in a similar manner; we model behaviours, not as functions, but as *relations*. (The conjunctions of equations we have been considering *are* relations.) We view the relation as imposing constraints (relating the signals on the various ports), which must be satisfied if the device is behaving properly, and use conjunction to model composition. We look in more detail at a simple example.

Transistor as Switch The MOS transistor has three ports, gate, source and drain, it is commonly used as a bidirectional switch. Informally, we

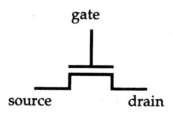

Figure 5.5: n-Transistor.

model its behaviour as a switch controlled by the signal on the gate as shown . (We shall see later that a simple switch model is not sufficient.) Although

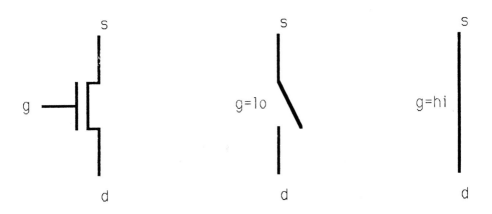

Figure 5.6: the switch model.

the gate can reasonably be regarded as an input, the source and drain can act as inputs or outputs. We cannot formalise this model as a function. For this example, we model signals as digital values, 0 or 1. There are eight possible combinations of signals on the ports. Only some of these are compatible with the switch model of the transistor. When the gate is 1 the switch is closed, and the signals on the source and drain must be the same. When the gate is 0, the signals on the source and drain are independent.

source		0	0	1	1
drain		0	1	1	0
gate	0	•	•	•	•
	1	•	×	•	×

Here, an × marks a combination which is disallowed. This diagram is analogous to the graphs of our earlier example.

Now consider connecting two transistors, $T1$, and $T2$. We can construct

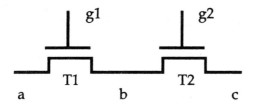

Figure 5.7: Connecting Two Transistors

a similar table - but we have five signals to consider.

		0	0	0	0	1	1	1	1
	b	0	0	1	1	1	0	0	1
	c	0	1	1	0	0	0	1	1
	00	•	•	•	•	•	•	•	•
$g1, g2$	01	•	×	•	×	×	•	×	•
	11	•	×	+	*	×	+	*	•
	10	•	•	+	+	•	+	+	•

Here, a + marks a combination which is disallowed by the behaviour of the first transistor, $T1$, an × marks a combination which is disallowed by the behaviour of $T2$, and an * marks a combination which is disallowed by both. The remaining points represent the behaviour of the system.

We can examine this table to analyse the behavior. For example, there are only two combinations allowed in which both $g1$ and $g2$ are 1, in both these cases, $a = c$. On the other hand, for any other pair of signals on the gates, because the signal on at least one gate is 0 any combination of signals on a and c is allowed (with a suitable signal at b).

If we choose to hide the connection, b, we see that the permissible patterns on the remaining ports are

		a	0	0	1	1
		c	0	1	1	0
	00		•	•	•	•
$g1, g2$	01		•	•	•	•
	11		•	*	•	*
	10		•	•	•	•

This is how we model structural abstraction (hiding internal connections); we see what combinations of signals on the visible ports are consistent with the behaviour of the consituent devices, for *some* signal on the hidden port.

Rather than tabulate (it should be clear that this method will not scale to much larger examples), we can argue informally about the behaviour of our simple circuit as follows: 'If the gate of a transistor is high, then the source equals the drain'. Suppose there is some signal b satisfying the constraints imposed by both $T1$ and $T2$. Because of $T1$, we see that, 'if $g1$ is high then a equals b'. Because of $T2$, we see that, 'if $g2$ is high then b equals c'. Combining these two statements, we see that, 'if $g1$ and $g2$ are both high, then a equals c'. Formal logic reduces such arguments to syntactic manipulation.

5.2.3   A Calculus of Behaviours

Reconsider our premise, 'If the gate of a transistor is high, then the source equals the drain'. This statement is *true* of any pattern of signals on the ports permitted by the behaviour. In fact, the behaviour is precisely the set of patterns for which the statement is true. The statement

$$\textbf{if } g = hi \textbf{ then } s = d$$

is related to the behaviour of the transistor in just the same way as, at a very different level of description, Ohm's law, $V = IR$, is related to the behaviour of a resistor. We will represent behaviours by statements; just as we use algebra to manipulate equations, we will use formal logic to manipulate these statements.

Formalism

> *'When I use a word,'* Humpty Dumpty said, in rather a scornful tone, 'it means just what I choose it to mean — neither more nor less.' 'The question is,' said Alice, 'whether you can make words mean so many different things.' 'The question is,' said Humpty Dumpty, 'which is to be master — that's all.'*
> Lewis Carroll, *Alice Through the Looking-Glass*

Relying on a specification in English, the designer and the client could put differing, but reasonable, interpretations on the specification and only discover the problem on delivery of the product. Only lawyers grow rich on such subtleties. How can we avoid them?

We follow Humpty Dumpty, and define the meaning explicitly. Once this is done we can abstract sound rules or equations for reasoning. Formal methods are characterised by rigorous treatment of syntax, semantics and proof. By formalizing, we remove the ambiguity inherent in English, or any other natural language.

Syntax and proof are formal in the sense that they can be represented in a computer program and manipulated without reference to the semantics. If they agree on a formal system, the parties to the contract could then stipulate that a proof checked by the computer program is to be the arbiter of correctness for an implementation.

We want a proof system for reasoning about behaviours. To do this, we first formalize the language (syntax) we will use and its meaning (semantics).

Syntax We start with a simple *language* which we will extend later. Our language has

	variables	$x, y, \ldots$
terms: $t, \sigma, \ldots$	constants	$a, b, \ldots$
	function applications	$f(t, \sigma)$

	equality	$\tau = \sigma$
	true	$\top$
formulae: $\varphi, \psi, \ldots$	conjunction	$\varphi \wedge \psi$
	implication	$\varphi \rightarrow \psi$
	hiding	$\exists x.\, \varphi$

Semantics   We will use formulae to express behaviours: variables will represent ports; constants, signal values; functions, primitive operations on signals; $\varphi \wedge \psi$ will represent the *conjunction* of behaviours φ and ψ; we use $\exists x.\, \varphi$ to represent the result of *hiding* x in the behaviour φ. Equality and implication will be used to construct primitive behaviours.

We now formalise our informal understanding of how behaviours compose. Let V be a non-empty set of signal values — at present we are using Boolean values. If P is the collection of *ports*, a pattern of signals at the ports is a function $v \in V^P$, we call such a function a *valuation*. A behaviour is a set of valuations — those consistent with the correct functioning of the device. Thus a behaviour is a subset of V^P, an element of $\mathcal{P}(V^P)$.

We assume each constant, a, is interpreted as $[\![a]\!] \in V$, and each function symbol, f, of the language is interpreted as an operation on values, $[\![f]\!] : V \times \ldots \times V \rightarrow V$. For each valuation $v \in V^P$, we extend v to define $v(\tau) \in V$, for each term τ, by setting $[\![f(\tau, \sigma)]\!] \stackrel{\text{def}}{\equiv} [\![f]\!]([\![\tau]\!], [\![\sigma]\!])$.

To each formula, φ, with variables $P = x_1 \ldots x_n$ we associate a behaviour, $[\![\varphi]\!]_P \subseteq v^P$ of a (possibly unrealizable) device with ports $x_1 \ldots x_n$[3].

$$[\![\top]\!]_P \stackrel{\text{def}}{\equiv} V^P$$

$$[\![\tau = \sigma]\!]_P \stackrel{\text{def}}{\equiv} \{v \in V^P \mid v\tau = v\sigma\}$$

$$[\![\varphi \rightarrow \psi]\!]_P \stackrel{\text{def}}{\equiv} \{v \in V^P \mid \text{if } v \in [\![\varphi]\!]_P \text{ then } v \in [\![\psi]\!]_P\}$$

$$[\![\varphi \wedge \psi]\!]_P \stackrel{\text{def}}{\equiv} [\![\varphi]\!]_P \cap [\![\psi]\!]_P$$

$$[\![\exists x.\, \varphi]\!]_P \stackrel{\text{def}}{\equiv} \{v \in V^P \mid \text{for some } w \in [\![\varphi]\!]_{P^\frown x}.\, w \uparrow P = v\}$$

$$[\![\varphi]\!]_{P^\frown x} \stackrel{\text{def}}{\equiv} \{w \mid w \uparrow P \in [\![\varphi]\!]_P\}$$

These definitions accomplish two things: First, the clauses for $=$ and $\rightarrow$ allow us to construct primitive behaviours. For example, our transistor behaviour is given by

$$[\![gate = hi \rightarrow source = drain]\!]$$

---

[3] Here, $\uparrow$ denotes function restriction; $P^\frown x$ is P together with a new port, x. Note that, by virtue of the final clause, $[\varphi]_P \in V^P$ is defined for any P which includes the visible ports. Hidden ports are 'out of scope'; their names may be reused without implying a connection.

and the behaviour of an **and** gate is given by

$$[\![c = a\&b]\!].$$

Second, the clauses for $\wedge$ and $\exists$ formalise our informal understanding of the composition of behaviours.

Rules We will want to analyse the relationship between specified and implemented behaviours — perhaps with a view to showing they are equal. The fundamental relationship we will attempt to axiomatise is that behaviours $A_1, \ldots, A_n$ together impose at least as much constraint as another behaviour B.

$$A_1, \ldots, A_n \vdash B \stackrel{\text{def}}{\equiv} [\![A_1]\!] \cap \ldots \cap [\![A_n]\!] \subseteq [\![B]\!]$$

Here are some rules[4] satisfied by this relationship[5]:

Structural Rules

Reflexive $\overline{A, \Gamma \vdash A}$

Monotone $\dfrac{\Gamma \vdash A}{B, \Gamma \vdash A}$

Transitive $\dfrac{\Gamma \vdash B \quad B, \Gamma \vdash C}{\Gamma \vdash C}$

These structural rules would suffice if hypotheses were represented as sets. However, if we view them as lists (as we might well in an implementation) we need the force of the following:

Duplicates $\dfrac{B, B, \Gamma \vdash A}{B, \Gamma \vdash A}$

Permutations $\dfrac{\Gamma \vdash A}{\pi(\Gamma) \vdash A}$ (where π is a permutation)

Propositional Rules

---

[4] A single line signifies that, if the relationship above the line holds then so does that below the line. A double line signifies that the relationships above and below the line are equivalent.

[5] We reassure the alert reader that the relationship is independent of the choice of P — Exercise!

True ($\top$) $\dfrac{}{\vdash \top}$

And ($\wedge$) $\dfrac{\Gamma \vdash A \quad \Gamma \vdash B}{\Gamma \vdash A \wedge B}$

Implies ($\rightarrow$) $\dfrac{A, \Gamma \vdash B}{\Gamma \vdash A \rightarrow B}$

Equality ($=$) $\dfrac{\Gamma[\tau/x] \vdash A[\tau/x]}{\tau = \sigma, \Gamma[\sigma/x] \vdash A[\sigma/x]}$ Equality implies indistinguishability[6].

Exists ($\exists$) $\dfrac{A, \Gamma \vdash B}{\exists x.A, \Gamma \vdash B}$ (where x is not a port in B or Γ)

The restriction on the rule for $\exists$ provides the correct scoping for hidden internal wires; we can only hide and expose an internal wire if its name does not clash with external wires with which it would otherwise be confused.

The reader will find it straightforward, if tedious, to check that these rules are valid for our semantics. This semantics is, in fact, a fragment of Tarski's semantics for predicate logic. The rules are well-known rules of logical inference [6].

Sequential Behaviour Adapting this model to encompass sequential behaviour is straightforward. We represent the behaviour of a device as a constraint on the *streams* of values at its ports. Streams are formalised as functions from natural numbers (representing discrete time ticks) to values. For example, a unit delay has the behaviour defined by:

$$\Delta(in, out) \stackrel{\text{def}}{\equiv} \forall t. \, (out \, (t+1) = in \, t).$$

Here, *in* and *out* are formal variables, typed as functions from natural numbers (representing discrete time ticks) to values. This representation of behaviours, as predicates on streams, has been widely exploited for formal verification; for example, by [2], [13], [18, 19, 20, 21], [15] and ourselves, [8, 9, 10].

---

[6] $A[\sigma/x]$ denotes the result of substituting σ for x in A. This substitution operation needs careful definition to ensure that variables in the expression being substituted are not confused with hidden variables, bound in the substitution context [6].

5.2.4 Verification Requirements

A behaviour expresses the combinations of signals allowable. A specification should express the acceptable behaviours. For example, the specification of an error correcting chipset could be

> *The chips implement boolean functions* $encode : 2^n \to 2^{n+m}$ *and* $decode : 2^{n+m} \to 2^n$ *such that, for any function* $error : 2^{n+m} \to 2^{n+m}$ *such that for all* $v \in 2^{n+m}$ *there is at most one i such that* $vi \neq (error\ v)i$ *we have*
>
> $$decode \circ error \circ encode = identity.$$

Specifications are higher-order objects — sets of behaviours. Methodologically, it is often more useful to express specifications at the same level as behaviours. We specify constraints which must be imposed, and allow any implementation which imposes at least these constraints. The verification condition is

$$implementation \vdash specification.$$

This picks out a set of acceptable behaviours, but not all possible sets of behaviours can be expressed in this way.

Because our logic is monotone (the transitivity of $\vdash$), this allows us to design hierarchically, basing a design on specified behaviour. On the basis of the specifications of our subcomponents, we can deduce that the design will impose certain constraints. If we use actual components that impose more constraints, then our design will still impose the required constraints (and will, typically, impose other, incidental constraints).

A typical specification is that for a unit delay:

$$\Delta(in, out) \overset{\text{def}}{\equiv} \forall t.\ out(t+1) = in\ t.$$

This allows an implementer to leave the output at time 0 undefined, or to initialise it to 0 or 1, and still satisfy the verification condition

$$\Delta\_Imp(in, out) \vdash \Delta(in, out).$$

Unfortunately, this verification condition is also satisfied by less satisfactory implementations. For example,

$$\bot \vdash \Delta(in, out).$$

Since, logically, $\bot \equiv (0 = 1)$, it is all too easy to implement *any* specification. Clearly our verification condition gives at most *partial* correctness — it is necessary but, as we have seen, not sufficient.

Many approaches to giving total correctness conditions have been proposed. We could insist that the implementation be *equivalent* to the specification — but we have seen how this solution foregoes many of the benefits of abstraction. [33] has proposed that the specification be divided into environmental constraints and behaviour, and that the verification condition be

$$constraints \vdash implementation \equiv behaviour.$$

In many cases, this provides the necessary flexibility of implementation without introducing the problem of 'false implies everything'. However, it does not accomodate examples, such as the unit delay, where we wish to allow flexibility at certain times, rather than at certain input signals. Another approach is to restrict our attention to directed devices — using a design style where each component is used directionally, and *design rules*[7] which ensure that the composition of components will not generate vacuous behaviours. For example, we may take the view that the intended behaviour is a function; the specification describes a relation within which the graph of the function must lie; formally, we must then make sure that our design rules only allow compositions that preserve the property (of behaviours) of including the graph of some function, and that the behaviours of our basic components have this property.

5.2.5   Abstractions

The need for abstraction arises as soon as our specifications are expressed at a level higher than the implementation. The specification stipulates a relation which must hold between abstract objects. These abstract objects will be represented by patterns of concrete data in space and time. If we let *Abs* be the mapping from concrete representations to the abstract objects they represent, the fundamental verification condition is

$$Implementation(\mathbf{x}) \vdash Specification(Abs(\mathbf{x})).$$

which is better stated as

$$y = Abs(\mathbf{x}), Implementation(\mathbf{x}) \vdash Specification(y).$$

since the abstraction used must be delivered along with the implementation if the device is to fulfill its intended use. (Note that there will usually be many abstract objects whose representation we must specify in this way. To

---

[7] Do not connect outputs together; do not create zero-delay loops; do not leave floating inputs.

have shewn more would only have obscured the basic idea behind yet more notation.)

We will later look in some detail at various examples of abstraction. Here, we make one simple, but subtle, point. In general, abstractions are *partial* (rather than total) functions; some patterns of concrete data may not correspond to *any* any abstract value. When we introduce such an abstraction, the resulting verification conditions contain terms which may not denote anything.

We could, alternatively, code these partial functions as relations, and write our verification condition as:

$$IsAbs(\mathbf{x}, y), Implementation(\mathbf{x}) \vdash Specification(y).$$

But we have anyway to take account of the fact that our abstract signals (sequences of abstract values, modelled as functions) may at some times be undefined; so we might as well accept the need to reason about partial functions and terms which may not denote. There is an extensive literature on this subject. Our approach follows [7]; we allow terms which may not denote, and introduce an 'existence predicate': $E\tau$ expresses the assertion that τ denotes[8] (something). We modify the quantifier rules to account for the fact that quantifiers range over existing things, and interpret equality to imply existence,

$$(\tau = \sigma) \vdash E\tau \wedge E\sigma;$$

and introduce a notion of equivalence,

$$\tau \equiv \sigma \overset{\text{def}}{\equiv} E\tau \vee E\sigma \rightarrow \tau = \sigma.$$

Equivalent terms are intersubstitutable in the logic.

5.2.6   Rules of Logic

We have represented behaviours as sets of assignments of signals to ports. We gave some simple rules for inclusion of behaviours earlier. In reasoning about behaviours, it is useful to consider other sets. Here we extend the rules to allow us a more expressive language. We extend the language we introduced earlier and use the general apparatus of predicate logic to reason about these sets. The semantics we gave earlier is just a fragment of Tarski's semantics for predicate logic.

Structural Rules   These are exactly as before.

---

[8] This assertion may be false.

Propositional Rules

We add the rest of the usual propositional connectives.

True ($\top$) $\overline{\vdash \top}$

And ($\land$) $\dfrac{\Gamma \vdash A \quad \Gamma \vdash B}{\Gamma \vdash A \land B}$

False ($\bot$) $\overline{\bot \vdash A}$

Or ($\lor$) $\dfrac{A, \Gamma \vdash C \quad B, \Gamma \vdash C}{A \lor B, \Gamma \vdash C}$

Implies ($\rightarrow$) $\dfrac{lrule A, \Gamma \vdash B}{\Gamma \vdash A \rightarrow B}$

Our logic is constructive, $\neg \varphi$ is expressed by $\varphi \rightarrow \bot$, and needs no further axioms or rules.

Identity, Existence and Quantifiers. Our logic is a so-called *free* logic in which terms are not presupposed to denote. Terms are descriptions rather than names. Identity , $\equiv$, and existence, E, are primitives. '$E\tau$', is to be interpreted as, 'τ describes something which exists'. Equality may be defined by, '$\tau = \sigma \iff E\tau \land E\sigma \land \tau \equiv \sigma$'.

Equivalence ($\equiv$) $\dfrac{\Gamma[\tau/x] \vdash A[\tau/x]}{\tau \equiv \sigma, \Gamma[\sigma/x] \vdash A[\sigma/x]}$
Identity is indistinguishability.

Extensional $\dfrac{E\tau, \Gamma \vdash \tau \equiv \sigma \quad E\sigma, \Gamma \vdash \tau \equiv \sigma}{\Gamma \vdash \tau \equiv \sigma}$
If two terms, in so far as they describe anything, are equivalent, then they are equivalent.

For All ($\forall$) $\dfrac{Ex, \Gamma \vdash A}{\Gamma \vdash \forall x.A}$ (where x is not free in Γ)

$$\frac{Ex, A, \Gamma \vdash B}{\exists x.A, \Gamma \vdash B}$$

Exists ($\exists$) $\quad \overline{\exists x.A, \Gamma \vdash B}$ (where x is not free in B or Γ)
Quantifiers range over things which exist.

5.2.7 Summary

The recent VLSI revolution is based on the MOS transistor. Using this as a primitive, we build gates, registers, counters, state machines, controllers, microprocessors and systems. We need to model and relate behaviours at many levels.

To model the behaviour of combinational logic, we model an output signal as a Boolean function of the signals on the inputs. Function composition models the connection of the output of one device to an input of another. Boolean algebra allows us to implement a given function quickly and accurately. To model general patterns of connection, and the hiding of internal wires, we need more-sophisticated models of behaviour.

We may not wish to label the ports as inputs and outputs. The classic example is a MOS transistor. The transistor is commonly used as a bidirectional switch. Informally, we model its behaviour as a switch controlled by the signal on the gate. (We shall see later that a simple switch model is not sufficient.) Although the gate can reasonably be regarded as an input, the source and drain can act as inputs or outputs. We cannot formalise this model as a function.

At higher levels of abstraction, we may want to allow the possibility that, for some inputs, the output is indeterminate. To represent a unit delay as a function, we have to say which logic value appears on the output at time zero. It is more realistic to say simply that all we know of the behaviour is that the value on the input at time t appears on the output at time t+1. This is not a functional relationship, the signal on the output is not determined by the signal on the input. However, it is constrained by it.

A solution, is to say that a device only allows certain combinations of signals on its ports. This representation of the behaviour as a subset of the possible combinations of signals is analogous to the representation of analogue behaviour of a power source by a curve showing which combinations of current and voltage it admits. The characteristics of a load are represented by a similar curve. To determine what happens if we connect the two, we intersect the two curves. We model the interconnection of digital devices in the same way.

The behaviour of a component is represented as a subset of the possible combinations of signals at its ports. When we compose components, their combined behaviour is given by intersecting their individual behaviours. Clearly, graphical or tabular methods do not provide a practical way of manipulating these subsets. However, formal logic provides a notation for describing and manipulating these models.

Predicate Logic provides a formal language for expressing relationships between things, together with a mathematical model of deductive reasoning about these relationships. Deductions are formal objects; their correctness can be mechanically checked. By modelling the notion of achievement of a specification by an implementation within predicate logic, we are able to divert the well-developed technology of computer-assisted formal reasoning to our ends.

Because the concrete signals we use to implement our circuits may, at times, not correspond to *any* abstract value, we must cater for reasoning about terms which may not denote anything.

We may compose the behaviour of a complex module from the behaviours of its parts, using logical operations to model connection, and hiding, of internal wires. A design may be verified by showing, using formal proof, that this behaviour entails the specification. Later, we will see that logic also allows us to express the abstractions that relate behaviours at differing levels of abstraction.

5.3   Goal-Directed Design

Software models allow the designer to experiment cheaply and flexibly. The CAD explosion of recent years is based on such modelling. Current CAD tools model geometry, structure and combinational behaviour. However, these models are not sufficient to support machine assistance for behavioural design at higher level of abstraction. We can only do this by formalising the designer's goals and the process of achievement of goals. This will allow us to build tools which coordinate, oversee and document design steps and maintain an agenda of goals which must be achieved to complete the design.

Theorem proving technology is now capable of specifying and verifying the behaviour of digital systems whose behaviour is so complex that exhaustive simulation is not feasible [Birtwistle 88]. However, theorem-proving methods are not accessible to the engineer, and are not integrated with existing CAD tools. Moreover, almost all published examples of verification are *post hoc*; verification only starts once the design is complete. In this section, we see how to integrate proof and design; starting from a specification, a proof of correctness is generated as a by-product of the design process. We start by presenting design as an example of goal-directed problem solving.

5.3.1 Goal-Directed Problem Solving

It is customary to contrast top-down and bottom-up design. We shall see
later that our formalism encompasses both. For the moment, we concentrate
on top-down design which we model informally as a goal-directed, problem-
solving activity. The goal-directed approach to problem solving involves *reducing* a goal to a list of subgoals which, if achieved, would allow us to achieve
the goal. Repeated reduction of goals to subgoals builds a tree (Figure 5.8).
Milner ([11]) formalised this process by introducing tactics and tacticals. A

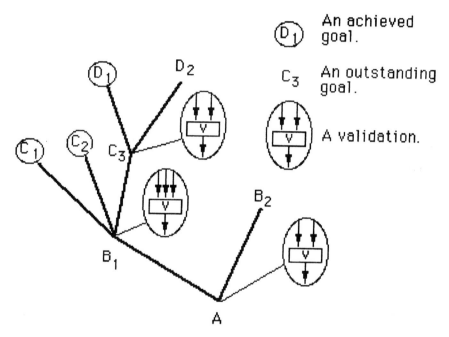

An achieved goal.

An outstanding goal.

A validation.

Figure 5.8: a prooftree.

tactic is applied to a goal to produce, not only the subgoals, but also for each
list $\{B_i\}$ of subgoals produced from a goal A, a way of combining achievements
of the $\{B_i\}$ to form an achievement of A. Such a 'way' is called a *validation*.
Using the notation of ML ([30])

- **type validation = achievement list → achievement**

- **type tactic = goal → (goal list * validation)**

Problem-solving generates a tree whose nodes are goals; the root of the tree is
the initial goal. As we construct the tree, using valid tactics, we also annotate

it with validations. When all the subgoals of a goal are achieved, the validation attached to it may be applied to achieve the goal. In the example, when D_2 is achieved, the tree of Figure 5.8 will collapse to Figure 5.9

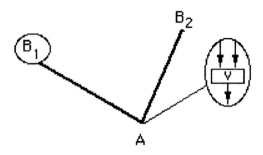

Figure 5.9: the reduced prooftree.

5.3.2   Proof-Search as Problem Solving

Milner used tactics to model goal-directed theorem-proving (see [28], where the originals of our diagrams are to be found). The goals in this domain are of the form *prove* φ, where φ is a formula. An achievement is a *proof*, or the *theorem* which is the last line of a proof. Primitive valid tactics correspond to the primitive *inference rules* of the logic. An inference rule may be viewed as a function from proofs to proofs. The validation of the tactic is the inference rule. For example, the rule $\dfrac{\vdash A \quad \vdash B}{\vdash A \wedge B}$ provides a function which given a proof of A and a proof of B yields a proof of $A \wedge B$. The corresponding tactic, CONJ_TAC, takes a goal which is a conjunction (it fails on all other goals) and produces its conjuncts as subgoals, together with the validation provided by the inference rule. Other tactics correspond to the other inference rules. This is all very well, but it would be excruciating to use only these tactics. Often we want to combine tactics algorithmically. (A simple example would be to repeatedly apply CONJ_TAC to all outstanding goals until all top-level conjunctions have disappeared.) Tactics may be combined using *tacticals*. New tactics may also be written. This raises a problem; how can we ensure that our tactics are correct? The user should only be allowed to apply validations to theorems, not to other objects, such as formulae, which look so like theorems that in a moment of misguided inspiration he may mistake one for the other. This is accomplished by a type discipline. Theorems form an abstract type, elements of which may only be constructed using the primitive inference rules. This ensures that, whatever tricks we may use in coding tactics and tacticals, we can never produce spurious theorems. The worst that can happen is that our validations fail to prove the required sequent.

5.3.3 Modelling the Design State

To apply these ideas to more general problems of design, we need to choose a sufficiently expressive language for modelling both initial and intermediate design goals; we have to keep track of the relationship between an original specification, a partial design, and the obligations which remain to be discharged in order to complete the design. Given a specification, the design task is to produce an implementation that *achieves* the specification. For which we write

$$implementation \Rightarrow specification.$$

However, the implementation cannot be described until the design is complete; so it cannot be specified in our initial goal. Our goal is thus refined as the design progresses. Our subgoals have the form:

$$partial\_design, pending\_decisions \Rightarrow remaining\_task$$

A tactic refining this goal would produce a validation which might make some of the pending decisions [16]. Although it is admirably general, this approach makes validations quite complex — essentially, they must code a bottom-up presentation of the design. We take a somewhat simpler approach. We treat the implementation as a variable which is progressively specialized (by instantiation) until the design is complete. Our subgoals have the form:

$$partial\_design, pending\_decisions \Rightarrow remaining\_task$$

Here the implementation has been specialized: *pending_decisions* is a variable which can accommodate further specialization; *partial_design* represents decisions taken so far (and may itself contain variables whose instantiation will represent further refinement of those decisions). The goal is to be understood informally as, 'find instantiations of the remaining variables such that the design task is achieved'. We can keep track of sufficient conditions (subgoals) for the correctness of the eventual design (goal). By this we mean that the goal follows logically from the subgoals. So the design state is represented by a *rule* (derived by composing primitive rules in some formal logic) of the form,

$$partial\_design, pending\_decisions \Rightarrow remaining\_task_1$$
$$\vdots$$
$$\frac{partial\_design, pending\_decisions \Rightarrow remaining\_task_n}{partial\_design, pending\_decisions \Rightarrow specification}$$

This should be read as saying, 'if pending_decisions, together with the partial design, accomplish the remaining tasks, then the conjunction of these decisions with the partial design will implement the entire specification'. This rule,representing the design state, is closely related to the more-general tree used earlier. The premises of the rule are the (unachieved) leaves of the tree; the conclusion of the rule represents (schematically, since uninstantiated variables may remain) what would be achieved at the root if the leaves were achieved. A rule is thus used to represent what has been achieved by a partial design; the relationship between the original specification, the partial design, and any remaining obligations which must be achieved to complete the design. If this rule is valid then so is any later specialization. Since variables, such as *pending_decisions*, express sharing, specializations made to discharge one sub-goal are automatically propagated to the others. We represent design steps as transformations which may be applied to produce valid rules from valid rules. By allowing variables in the initial specification we can also accommodate refinement of the specification within the paradigm. The system ensures that only logically valid rules are generated. This approach has its analogue in theorem proving in ISABELLE [32], which borrows Milner's ideas of tactical theorem proving, but manipulates derived rules, rather than theorems. For our purposes this is a much simpler approach, because it allows tactics that refine the goal by instantiation to be implemented straightforwardly. To summarize: The system's representation of the state of a design is always a valid rule. A new rule is produced with each refinement made by the designer. Refinements transform rules by changing the 'partial_design' part of the bottom line, as well as totally reforming the top line. That is, the rule above changes to a rule of the same form, but with a simpler set of remaining tasks and a refined design:

$$\frac{refined\_design, pending\_decisions \Rightarrow new\_task_1}{refined\_design, pending\_decisions \Rightarrow new\_task_n}$$
$$refined\_design, pending\_decisions \Rightarrow specified\_behaviour$$

At the start of the design, there is a complete specification and an entirely free implementation. This is represented by the rule,

$$\frac{pending\_decisions \Rightarrow specified\_behaviour}{pending\_decisions \Rightarrow specified\_behaviour}$$

This starting point is the tautology, 'if I accomplish the goal, then I will have accomplished the goal'. No design decisions have been made: pending_decisions is a variable representing a list of pending decisions. We then incrementally transform the rule representing our design, starting with this tautology and ending with a rule with no conditions. The final rule will have the form

$$\frac{}{completedesign \;\Rightarrow\; Specification} \; .$$

At this stage, there are no more remaining tasks, and the design is complete. To implement this paradigm we need to formalise specifications, implementations and the logic of achievement.

5.3.4 Formalising Design Refinement

Predicate Logic provides a formal language for expressing relationships between things, together with a mathematical model of deductive reasoning about these relationships. Deductions are formal objects; their correctness can be mechanically checked. By modelling the notion of achievement of a specification by an implementation within predicate logic, we are able to divert the well-developed technology of *computer-assisted formal reasoning* to our ends. This provides a new approach to the problems of high-level synthesis and design assistance.

We use relations to represent behaviour, and sequents to represent the validation task as before. The fusion of verification with design is made possible by formally representing the relationship between three things:

- the ultimate goal

- what has already been done

- the outstanding tasks.

This relationship may be represented as a *rule* of logic. Formal logic provides a basis for representing intermediate design states. Proof tactics in a refinement-based theorem prover can be used to formalise design strategies. Refinement-based proof, as exemplified by ISABELLE ([32]) and LAMBDA ([1]) inherits many ideas (notably, the use of an ML's type discipline to construct a safe type, and a higher-order tactical approach to goal-directed proof) from LCF and HOL ([11, 12, 14]), but explicitly allows proof to proceed by specialisation of the goal. This is crucial to our representation of the design process.

LAMBDA and ISABELLE are formal proof systems which manipulate objects called *rules*. A rule is a member of a *safe type* of the system, meaning that all rules produced are *sound*. A rule consists of the following parts :

$$Premise_n$$
$$\vdots$$
$$\frac{Premise_1}{Conclusion}$$

Each of the *premises* and the *conclusion* is a *sequent* of the form :

$$hypotheses \vdash assertion$$

For a sequent to *hold*, the *assertion* must be TRUE whenever all the *hypotheses* are TRUE. For the above rule to be sound, the conclusion must hold whenever all of the premises hold.

The current state of a design can be represented by a rule. Before any work has been done, no implementation has been defined and the rule is :

$$\frac{UnknownImp \vdash Spec}{UnknownImp \vdash Spec}$$

This states the tautology that if *UnknownImp* implements *Spec* then *UnknownImp* implements *Spec*. *UnknownImp* is a variable representing what is left to be synthesised. As the design proceeds we define some of *UnknownImp* and in doing so we produce simpler specifications, $ReducedSpec_i$ to be implemented :

$$\frac{\begin{array}{c} PartialImp + UnknownImp_1 \vdash ReducedSpec_n \\ \vdots \\ PartialImp + UnknownImp_1 \vdash ReducedSpec_1 \end{array}}{PartialImp + UnknownImp_1 \vdash Spec}$$

Note that partitioning has occurred — each premise can be implemented separately. *PartialImp* is normally a list of components. In simple cases, at the end of the synthesis there will be no premises remaining in which case the design is complete :

$$\frac{}{CompleteImp \vdash Spec}$$

Often, however, there are some constraints remaining such as maximum input range and microcode Specifications; such constraints represent operational parameters of the design :

$$\frac{}{Constraints, CompleteImp \vdash Spec}$$

Rules can be combined to produce a new rule by *resolution*; the conclusion of one rule is *unified* with the first premise of another to produce a set of *bindings* called an *environment*. The resulting rule is obtained from the first rule by replacing its first premise by the premises of the second rule, and *pruning* in the environment all the sequents of the resulting rule (i.e. any variables in the sequents which have bindings in the environment are replaced by the values of those bindings). Resolution can be used to modify a rule representing the current design state; the new rule produced by combining this rule with a rule representing a design decision, is a representation of a new design state. Because only sound rules can be produced, the resulting design is correct by construction at the behavioural level.

Unification is the process of finding an environment in which two expressions are equivalent. For example, $x + fy$ and $ga + b$ are equivalent in an environment with bindings $x := g\,a$ and $b := f\,y$. In this environment, both expressions prune to $ga + fy$. In this example $x\ y\ a\ b\ f$ and g are first-order (meta-) variables - they can be bound to expressions; LAMBDA also has second-order (meta-) variables such as $\mathcal{P}$ in the expression $\mathcal{P}(x, y)$. $\mathcal{P}$ is a *syntactic function* describing a method of using x and y to create an expression; such functions are created on-the-fly by higher-order unification. Such variables allow us to represent[9] *rule schemas* such as the commutativity of $+$:

$$\frac{hypotheses \vdash \mathcal{P}(b + a)}{hypotheses \vdash \mathcal{P}(a + b)}$$

This rule enables the operands of a $+$ to be commuted. Often there will be more than one $+$ and so we may wish to specify which ones we want to affect. In the LAMBDA system, this may be done using a *guide*, an annotation of the abstract syntax tree of the expression, obtained by selecting the appropriate occurrence of $a + b$ using a mouse. This enables the unification to produce the required result (or *unifier*) first in the stream of results (while allowing other unifiers to be produced by *backtracking*).

Resolution is only one way of transforming a rule; a more general method is the use of *tactics*. A tactic (another safe type in LAMBDA) transforms a premise into a stream of $(premiselist, environment)$ pairs. A new rule is created from an existing one by transforming a premise of the old rule, taking one of the elements $(prems, env)$ of the resulting stream, replacing the premise in the old rule by $prems$, then pruning (replacing variables by the expressions to which they are bound) the whole rule in env.

Design steps are carried out until some or all of the specification has been synthesised. A design step typically consists of bringing one or more components into the schematic, possibly making some connections, and then selecting an option from the operations menu.

Certain design steps allowing a specification to be simplified could lead to invalid circuits. For example, we have seen that our model fails if outputs are connected together. The underlying behavioural representation has no notion

---

[9] ISABELLE achieves a similar result using higher-order unification.

of *outputs* or even *components* so we have to filter out any results which, when interpreted as circuits, represent a violation of a check such as this. Other necessary checks include zero-delay loop detection. Theoretically, we have a notion of valid circuit, and a semantic function which generates the behaviour (predicate) associated with a valid circuit. In practice, it is easier to work, and apply design-rule checks, at the level of behaviours.

Introducing a Delay To give a first example of a design step, we recall the definition of a unit delay:

$$\Delta(in, out) \overset{\text{def}}{\equiv} \forall t \, (out(t + 1) = in \, t)$$

Adopting this definition, we may derive (in predicate logic) the following rule schema:

delayIntro
$$\frac{\Delta(in, out), \Gamma \vdash \mathcal{P}[in \, t]}{\Delta(in, out), \Gamma \vdash \mathcal{P}[out(t + 1)]}$$

Informally, this rule should be read backwards. It says, "if you want to achieve some property of $out(t+1)$ it suffices to achieve the same property of in t provided *in* and *out* are connected by a unit delay". (Formally, $\mathcal{P}$ is a second-order metavariable ranging over contexts.) Given some goal (occurring as a premise of the rule representing our proof state) that we wish to achieve, we can apply this rule by unifying the goal with the conclusion of the delayIntro rule, and then replacing the goal with the premise of the delayIntro rule. Typically, the unification will have the effect of introducing a delay component into our design, and connecting its output to a terminal on which it is required to place a particular value at some time $t + 1$. This requirement will then be replaced by the requirement to place the same value on the input of the delay at time t.

5.3.5 Example

We give an example of a (mildly) non-trivial design. The initial goal is to multiply four numbers:

$$result = (in1 * in2) * (in3 * in4)$$

We have at our disposal a multiplier with the following behaviour

$$MUL(a, b, out) \overset{\text{def}}{\equiv} \forall t \, (out(t + 4) = a \, t * b \, t)$$

from which we derive the following rule

MULIntro

$$MUL(a, b, out), \Gamma \vdash a\, t = x$$
$$MUL(a, b, out), \Gamma \vdash b\, t = y$$
$$\underline{MUL(a, b, out), \Gamma \vdash \mathcal{P}[out(t + 4)]}$$
$$MUL(a, b, out), \Gamma \vdash \mathcal{P}[x * y]$$

Applying this rule to the goal,

$$\Gamma \vdash result = (in1 * in2) * (in3 * in4)$$

(with x unified with $(in1 * in2)$ and y with $(in3 * in4)$) introduces a multiplier, $MUL(a, b, out)$, to the design and leaves us with the goals

$$MUL(a, b, out), \Gamma \vdash a\, t = (in1 * in2)$$
$$MUL(a, b, out), \Gamma \vdash b\, t = (in3 * in4)$$
$$MUL(a, b, out), \Gamma \vdash result = out(t + 4)$$

Applying the rule twice more, using the same multiplier, we arrive at the goals

$$MUL(a, b, out), \Gamma \vdash a\, t = (out(t1 + 4))$$
$$MUL(a, b, out), \Gamma \vdash b\, t = (out(t2 + 4))$$
$$MUL(a, b, out), \Gamma \vdash a\, t1 = in1$$
$$MUL(a, b, out), \Gamma \vdash b\, t1 = in2$$
$$MUL(a, b, out), \Gamma \vdash a\, t2 = in3$$
$$MUL(a, b, out), \Gamma \vdash b\, t2 = in4$$
$$MUL(a, b, out), \Gamma \vdash result = out(t + 4)$$

This represents a design decision — that we will use one multiplier — and leaves open how we will schedule the operations and complete the design. We can effect a schedule by introducing a common time reference, s, instantiating $t1$ to s, $t2$ to $s + 1$ and t to $s + 5$:

$$MUL(a, b, out), \Gamma \vdash a(s + 5) = (out(s + 4))$$
$$MUL(a, b, out), \Gamma \vdash b(s + 5) = (out(s + 5))$$
$$MUL(a, b, out), \Gamma \vdash a\, s = in1$$
$$MUL(a, b, out), \Gamma \vdash b\, s = in2$$
$$MUL(a, b, out), \Gamma \vdash a(s + 1) = in3$$
$$MUL(a, b, out), \Gamma \vdash b(s + 1) = in4$$
$$MUL(a, b, out), \Gamma \vdash result = out(s + 9)$$

This raises the interesting question of conflicting requirements; we have to get values from different sources on to a and b at different times. Logically, we could unify a with out (corresponding to making a physical connection) to discharge the first of these subgoals. However, when we put the value $in1$ on a (at time s), to discharge the third goal, we might be in conflict with the value on *out*. The logical model of this situation (falsity) does not make physical sense, and we have to impose design rules (in this case that outputs of subcomponents should not be connected together or to primary inputs) to ensure that our formal model reflects reality. We can proceed with the design by introducing multiplexors on the inputs. A multiplexor has the behaviour

$$MUX(c, a, b, out) \stackrel{\text{def}}{\equiv} \forall t\, (out\ t = if\ c\ t\ then\ a\ t\ else\ b\ t)$$

From which we derive the rules

MUXTrue
$$\frac{MUX(c, a, b, out), \Gamma \vdash c\,t = true \qquad MUX(c, a, b, out), \Gamma \vdash \mathcal{P}[a\,t]}{MUX(c, a, b, out), \Gamma \vdash \mathcal{P}[out\ t]}$$

MUXFalse
$$\frac{MUX(c, a, b, out), \Gamma \vdash c\,t = false \qquad MUX(c, a, b, out), \Gamma \vdash \mathcal{P}[b\,t]}{MUX(c, a, b, out), \Gamma \vdash \mathcal{P}[out\ t]}$$

Applying these judiciously, we introduce the components

$$MUX(c, a1, a2, a), MUX(c, b1, b2, b), MUL(a, b, out)$$

and arrive at the goals

$$
\begin{aligned}
&\ldots \vdash c(s+5) = true\\
&\ldots \vdash c(s+1) = false\\
&\ldots \vdash c\,s = false\\
&\ldots \vdash a1(s+5) = (out(s+4))\\
&\ldots \vdash b1(s+5) = (out(s+5))\\
&\ldots \vdash a2\,s = in1\\
&\ldots \vdash b2\,s = in2\\
&\ldots \vdash a2(s+1) = in3\\
&\ldots \vdash b2(s+1) = in4\\
&\ldots \vdash result = out(s+9)
\end{aligned}
$$

where we have elided repeated mention of the components. The first three of these express the control requirements of the multiplexors, the fourth can be modified using the rule delayIntro to arrive at the components

$$\Delta(d, a1), MUX(c, a1, a2, a), MUX(c, b1, b2, b), MUL(a, b, out)$$

and goals

$$... \vdash c(s + 5) = true$$
$$... \vdash c(s + 1) = false$$
$$... \vdash c\, s = false$$
$$... \vdash d(s + 4) = (out(s + 4))$$
$$... \vdash b1(s + 5) = (out(s + 5))$$
$$... \vdash a2\, s = in1$$
$$... \vdash b2\, s = in2$$
$$... \vdash a2(s + 1) = in3$$
$$... \vdash b2(s + 1) = in4$$
$$... \vdash result = out(s + 9)$$

Finally, we connect (unify) d and b1 to out to arrive at the design

$$\Delta(out, a1), MUX(c, a1, a2, a), MUX(c, out, b2, b), MUL(a, b, out)$$

This makes two of our goals trivial (by reflexivity of equality) and leaves

control specification
$$... \vdash c(s + 5) = true$$
$$... \vdash c(s + 1) = false$$
$$... \vdash c\, s = false$$

input specification
$$... \vdash a2\, s = in1$$
$$... \vdash b2\, s = in2$$
$$... \vdash a2(s + 1) = in3$$
$$... \vdash b2(s + 1) = in4$$

output specification $$... \vdash result = out(s + 9)$$

This simple example shows how a few simple design steps may be modelled within our paradigm. It also shows the need for machine assistance. By writing tactics and tacticals it is possible to model more sophisticated design steps, and present them to the designer as atomic design actions. Forward proof also plays a role within our paradigm, in modelling bottom-up design and design optimisation.

5.4   Abstraction

Modelling abstraction is an integral part of generating verification conditions. We cannot *in abstracto* verify that a particular piece of hardware implements a given specification; we can only verify that it does so *modulo* certain abstractions. No chip multiplies two numbers, but a lot of money has been made from chips which when presented with particular representations of numbers as digital signals, produce, after some delay, a representation of their product.

There is usually no verification of the correctness of abstractions used in refining a design to the point where it can be mapped to hardware. Even more critical is the lack of documentation of abstractions and the design management problems this causes. (For example what assumptions have been made about arithmetic of integers within finite word-lengths or even worse what abstractions are assumed in the simulation algorithms.)

We recall the general verification condition

$$y = Abs(\mathbf{x}), \; Implementation(\mathbf{x}) \vdash Specification(y).$$

There are infinitely many different abstraction functions we could use. However, many are common enough to be considered clichés, and some systematic study is possible. For example, an integer may be represented in parallel as the pattern of bits on a bus at a time t, or in serial as the bits on a wire at some sequence of times. It is possible to combine temporal and spatial representations — for example, presenting a 16 bit number as two bytes presented at different times on a single 8-bit buss. In general, we must explain how the high-level data of the specification is represented in terms of lower levels. It is useful, but often impossible, to factor this into separate timing and data abstractions, as is done by [26].

5.4.1   Data Abstraction

Specifications may use abstract datatypes, such as 'instruction' or 'integer'; implementations use bits. Data abstractions relate the abstract data to its concrete representation. A simple, and common form of data abstraction is to view a bunch of wires as a buss. The bunch of wires is formally represented as a tuple (or perhaps list) of sequences of values; the buss as a single sequence of tuples of values given by

$$buss \; t \; = \; < buss_1 \; t, \dots buss_n \; t >$$

More generally, a *pure* data abstraction is given by a *total* function

$$f : A_1 \times \cdots \times A_n \to B$$

(in our example, this function was tupling). The verification requirement is simple to express ([26]):

$$Implementation(x_1, \ldots, x_n) \vdash Specification(f \circ < x_1, \ldots x_n >).$$

Here, angle brackets indicate tupling of functions:

$$< x_1, \ldots x_n > = \ \mathbf{fn}t => (x_1 \ t, \ldots x_n \ t)$$

Equivalently, we may formulate this by saying that to the concrete tuple of sequences

$$\mathbf{x} = < x_i >_{i \in n}$$

there corresponds the abstract sequence

$$Abs(\mathbf{x}) \stackrel{\mathrm{def}}{\equiv} f \circ < x_1, \ldots x_n >$$

Bits are themselves an abstraction; they make a good example of a necessarily *partial* abstraction function. In MOS combinational logic, we assign the value **hi** to voltages above some threshold (typically $3V$), and the value **lo** to voltages below some threshold (typically $1V$)[10]. When we say that an inverter inverts, we mean that *for input voltages which correspond to well-defined logic values*, the logic value corresponding to the output voltage is well-defined, and is the inverse of the value corresponding to the input.

Looking at this example, we have a concrete domain (voltage) and an abstract domain (boolean), related by a *partial* function

$$abs : voltage \rightharpoonup boolean.$$

The concrete behaviour (given by the graph of voltage against voltage) implements the abstract specification of the inverter in the sense that

$$\forall V_{in}, V_{out}. \ inv\_imp(V_{in}, V_{out}) \wedge abs(V_{in}) \in bool \rightarrow$$
$$abs(V_{out}) \in bool \wedge inv\_spec(abs \ V_{in}, abs \ V_{out}).$$

It is noteworthy that, although the abstract specification is symmetric, this requirement introduces an asymmetry between input and output at the implementation level.

When we argue at the abstract level, we must always bear in mind that the abstraction function may be partial. Consider, for example, the simple circuit in Figure 5.11. Engineers will be familiar with the timing problems it

---

[10] [24] identify these two thresholds. This places unrealistic demands on the gate designer. (Exercise — Why?)

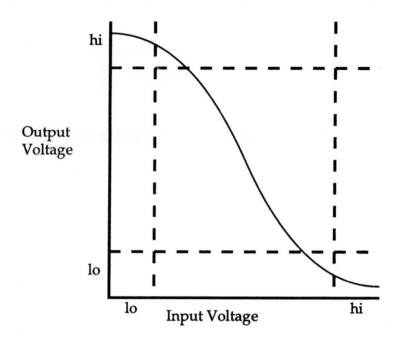

Figure 5.10: Inverter Characteristic

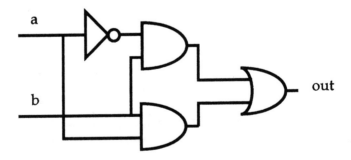

Figure 5.11:

can cause[11]; it can also introduce logical errors. We suppose that each gate performs its allotted boolean function. By trivial boolean reasoning we see that the behaviour will be *out* = *b*. However, changing the input *a* can give rise to 'incorrect' outputs because of features of the transfer functions of the various gates at the concrete level. Consider Figure 5.12, which shows the underlying behaviour of an inverter, and of an **and** gate with one input tied *hi*. If, in Figure 5.11 we tie *b hi* and apply the voltage *V* to *a*. Then the output of the inverter will be a logic *lo*, so the output of the upper **and** gate will be *lo*; the output of the lower **and** gate will also be *lo*. The output will be a logic *lo*, contrary to our analysis at the boolean level.

Fortunately, many data abstractions are better behaved than this. When the abstraction function is total life is much simpler: we may use the simpler verification condition

$$imp(in, out) \vdash spec(abs\ in, abs\ out).$$

5.4.2 Temporal Abstraction

Time is complex, partly because we have many different models of time, partly because the individual models may be complex, and, mostly, because we need to relate the different models. To analyse continuous change, it is usual to model time using real numbers. When modelling digital behaviour, we may be interested in the sequence of states of a system, or we may require more detailed models of time which allow us, for example, to talk about setup and hold times, or concurrent systems.

Temporal abstractions are common throughout digital design; an assembly code view of a microprocessor may abstract from the realities of prefetch, pipelining and microinstructions, and view each instruction as atomic. Each time-step at this level is implemented in terms of a (variable) number of microinstructions. Each microinstruction step may, in turn, be implemented in some number of system clock cycles. Finally, various submodules may be clocked at multiples of the system clock frequency. (Of course, we could continue further in the other direction and consider the compilation of a statement in a high-level language to a number of machine instructions, and so on.) Timing abstractions are needed (in addition to data abstractions) to relate these levels.

Simple temporal abstraction maps each abstract time *t* to a concrete time *ft*. For example, slow-down ([23]) can be modelled by the abstraction function

$$ttfnt => n \times t.$$

---

[11] Changing the input *a* can give rise to transient 'incorrect' outputs (glitches) because of unequal delays in propagating the effects of the change.

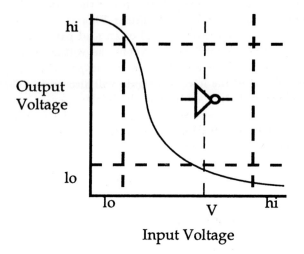

Input Voltage

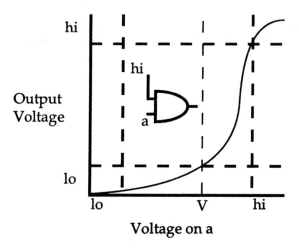

Voltage on a

Figure 5.12:

the abstract sequence of values consists of every n^{th} value of the concrete sequenbstractionmaythenbeappliedtothesevalues.Thetemporal
abstraction of a signal is again given by composition (this time on the other side):

$$Abs(x) \stackrel{\text{def}}{\equiv} x \circ f.$$

For even a simple, non-pipelined, microprocessor the mapping is more complicated — we typically define the sequence of abstract states by sampling the microstate at those times when the microprogram counter is zero, and then applying a data abstraction. This might be formalised as follows:

```
fun next p a t = if p (a t) then t else next p a (t + 1) ;
fun start state = mupc(state) = zero ;
fun timeAbs run 0 = next start run 0
|   timeAbs run (n + 1) = next start run (timeAbs run n) ;
val Abs run = stateAbs o run o (timeAbs run) ;
```

In general, different abstract signals in a system may be concretely represented via different timing abstractions. Furthermore, data and timing abstractions may be combined: a stream of bytes, $< b_i >_{1 \leq i \leq 8}$, each represented bit-serially, with a single bit separating successive bytes, is represented by the abstraction

$$((Abs(x))t)_i \stackrel{\text{def}}{\equiv} x(9t + i).$$

More sophisticated abstractions, again based on a monotone mapping f as above, are common. For example, the notion of a button being pressed at a user-level 'time' t has many variations:

- the corresponding signal is high at at least one system-level 'tick', s with $f(t) \leq s < f(t+1)$

- the signal rises at some time in the interval $[f(t), f(t+1))$

- the signal is high throughout the interval

Each of these alternatives is itself subject to any number of variations — we might ignore the first n ticks in each interval, and so on. Or we might view the system completely differently; perhaps, the user-level view of the same system counts the number of times the button is pressed and views the button as a sequence of numbers.

Exercises

1. A 60Hz boolean signal represents the sequence of states of a mouse button. At the user-level, this signal is abstracted to form a sequence of 'mouse events' $\in$ {Click, DoubleClick, NoClick}. Formalise a suitable abstraction function.

2. Consider the following informal specification for a design exercise:

> Design a digital stopwatch with a three digit, seven-segment display (to read seconds, tenths of seconds, and tens of seconds, and two control buttons, 'reset' and 'start/stop'. When the reset button is pressed the display is cleared. The start/stop button is used to start and stop the clock
>
> You are provided with a 1MHz system clock, and a 10Hz digital signal synchronised with the system clock. You may assume that the buttons also produce synchronised digital signals.

Your present tasks are as follows:

- Formalise the specification at the user level (10Hz clock).
- Write abstraction functions to express the controls and display in terms of 1MHz signals.
- Write the resulting verification requirement for an implementation.

5.4.3 Eliminating Abstractions

In top-down design, we aim to design a system in terms of abstract components which we will later implement. If our abstract components are specified in terms of abstract data, then their implementations will include, for signals which are internal to the system, abstraction specifications which we need to eliminate.

Suppose we have designed a system at an abstract level, and proved

$$\exists b.\ ASpec(a, b) \wedge BSpec(b, c) \vdash SysSpec(a, c).$$

If we now implement $ASpec$ and $BSpec$ at a concrete level, we have

$$a = aAbs(\mathbf{x}), b = bAbs(\mathbf{y}), AImp(\mathbf{x}, \mathbf{y}) \vdash ASpec(a, b)$$

and

$$b = bAbs(\mathbf{y}), c = cAbs(\mathbf{z}), BImp(\mathbf{y}, \mathbf{z}) \vdash BSpec(b, c).$$

We want to show that

$$a = aAbs(\mathbf{x}), c = cAbs(\mathbf{z}), \exists \mathbf{y}.\ AImp(\mathbf{x}, \mathbf{y}) \wedge BImp(\mathbf{y}, \mathbf{z}) \vdash SysSpec(a, c).$$

A sufficient condition is

$$a = aAbs(\mathbf{x}), AImp(\mathbf{x}, \mathbf{y}) \vdash \exists b.\ b = bAbs(\mathbf{y}).$$

This can be expressed more concisely with the E predicate:

$$E(aAbs(\mathbf{x})), AImp(\mathbf{x}, \mathbf{y}) \vdash E(bAbs(\mathbf{y})).$$

5.5 Top-Down Design

The inherent complexity of integrated circuits creates problems when verification is attempted on completed designs; these are alleviated by integration of design and verification throughout the design process. The fusion of verification with design is made possible by formally representing the relationship between three things: the ultimate goal, what has already been done, and the outstanding tasks. We have described how this relationship may be represented as a rule of logic, and manipulated by a suitably designed proof assistant. Starting from a specification, the proof of correctness may be generated as a by-product of the design process.

Sequences of interactions with the designer can be replaced by interactions generated by the machine — algorithmically or heuristically. We also have the opportunity of providing guidance for the designer which goes beyond behavioural correctness. We conclude by running through a simple design example, where abstractions are introduced in the course of formally-based design. Future work will build on this approach to produce an 'Intelligent Design Assistant' capable of design synthesis.

5.5.1 Example — an IIR Filter

We consider an infinite input response filter; this example is of particular interest because it uses recursion. The specification states that the output at time $t + 1$ is given by a constant multiple of the input at time t added to another constant multiple of the output at time t. Thus we have

$$IIR\ 0 = 0$$

(to initialise) and

$$IIR\ (t + 1) = a * input\ t + b * IIR\ t$$

where a and b are constants and the signal input is represented by a function from time to signal value. We can express this formally by defining the function IIR:

```
fun IIR input a b 0       =  zero
 |   IIR input a b (t + 1) =  (a * (input t))
                             + (b * (IIR input a b t));
```

The syntax used for the function definition is that of ML [30].

The object of the synthesis exercise will be to achieve an implementation which satisfies the specification i.e.

$$IMPLEMENTATION \vdash \forall t.\ output(t + m) \equiv IIR\ input\ a\ b\ t$$

If we have an input and an output and an implementation then for all times t, the output at time $t + m$ is *IIR input a b t*. Note that we have left unspecified the latency m of the system; this will be generated during the synthesis.

Implementation We choose to implement the IIR described above by using pipelined adders and multipliers with 2 and 4 unit delays respectively. The definition of an ADD is just

$$ADD(x, y, z) \stackrel{\text{def}}{\equiv} \forall t.\ z\,(t + 2) \equiv x\,t + y\,t$$

The internal time scale of the ADD is not the same as that of the specification for the IIR, so we define an abstraction to express the fact that the external signals are clocked n times more slowly than the internal signals

$$MACRO(n, x, y) \stackrel{\text{def}}{\equiv} (y \equiv \mathbf{fn}\ t => x\,(n * t))$$

$$MICRO(n, x, y) \stackrel{\text{def}}{\equiv} (x \equiv \mathbf{fn}\ t => y\,(n * t))$$

These abstractions can now be used to relate the external and internal signals. Our starting point is the rule

$$\frac{\begin{array}{l} INPUT(input),\ MICRO(n, input, x), \\ MACRO(n, y, output),\ OUTPUT(output) \\ \qquad\qquad\qquad \vdash \forall t.\ output(t + m) \equiv IIR\ input\ a\ b\ t \end{array}}{\begin{array}{l} INPUT(input),\ MICRO(n, input, x), \\ MACRO(n, y, output),\ OUTPUT(output) \\ \qquad\qquad\qquad \vdash \forall t.\ output(t + m) \equiv IIR\ input\ a\ b\ t \end{array}}$$

The object of the synthesis is to derive an implementation which achieves this specification. Values for both latency m and clock ratio n are determined during the design process. We must minimise n so as to get as fast a result as possible with respect to the internal clock rate.

Before introducing components to fulfill the specification we must first expand the definition of the IIR and convert the specification into a form which depends on the internal wires x and y. The recursive nature of the definition of the IIR leads naturally to a proof which uses induction on time, t. First using the definition of IIR at time $t + 1$ in terms of that at t, and then using the inductive hypothesis to substitute *output* at time $t + m$ for IIR at time t, we rewrite the requirements as follows:

$$output\ m\ =\ IIR\ input\ a\ b\ 0$$
$$=\ zero$$

$$output((t+1)+m)\ =\ IIR\ input\ a\ b\ (t+1)$$
$$=\ a*(input\ t)+b*(IIR\ input\ a\ b\ t)$$
$$=\ a*(input\ t)+b*output(t+m)$$

We can then use the relation between external and internal clock speeds to convert these conditions on *input* and *output* into conditions on x and y.

$$y(n*m)\ =\ zero$$
$$y(n*(t+1+m))\ =\ a*x(n*t)+b*y(n*(t+m))$$

Now use the ADD and MUL components to implement the operations. We allow a component to be used pipelined for more than one operation, so an addition is achieved by assigning the two values to be added to the inputs of the ADD at a time $t1$, and then replacing their sum by the output of the ADD at time $t1+2$. The procedure used is to add components until all the additions and multiplications have been implemented, and then to assign times to the various operations. The times are assigned using a *just in time* scheduling algorithm, which attempts to maximise the speed of the design by only performing operations just before their results are needed. During the course of this scheduling, values for m and n are calculated. They will depend on the way in which the ADDs and MULs are used. If one unfolding of the recursion is taken as the basis for synthesis, the use of two MULs generates a clock ratio $n=6$, while the use of one MUL gives a slower design with $n=7$.

We can however achieve a much faster design by unfolding the definition of IIR at time $t+2$ in terms of IIR at time t. We rewrite the specification as

$$IIR\ input\ a\ b\ 0\ =\ zero$$
$$IIR\ input\ a\ b\ 1\ =\ a*input\ 0$$
$$IIR\ input\ a\ b\ (t+2)\ =\ a*input(t+1)$$
$$+\ b*(a*input\ t+b*IIR\ input\ a\ b\ t)$$

This formulation does not in itself yield a faster implementation. The critical path which results for x and y is no different from that we had earlier, and so cannot be traversed more quickly. However, the user is now free to try a number of different approaches to rearrange the specification. An obvious one is to use the facts that addition and multiplication are associative and distributive to rebracket the above :

$$IIR\ input\ a\ b\ 0\ =\ zero$$
$$IIR\ input\ a\ b\ 1\ =\ a*input\ 0$$
$$IIR\ input\ a\ b\ (t+2)\ =\ a*input(t+1)$$
$$+\ (b*a)*input\ t+b^2*IIR\ input\ a\ b\ t$$

Now the multiplication of the constants can be done statically, and a speed-up can be achieved.

Once scheduling has been performed, the final design step uses a pre-programmed algorithm which automatically introduces multiplexers and delays so as to connect the arithmetic components together and fulfill the various conditions on their input values. This leads to timing conditions on the control signals of the multiplexers, which are left in the final design state. There is no need to 'prove' these; they just specify the multiplexers' control requirements. In general, many blocks being synthesised at one level may generate such requirements. The requirements generated in this way form part of the specification at the next level, and may be considered together or re-partitioned as appropriate. Partitioning at one level of abstraction does not force partitioning of the resulting control requirements at the next level. As a result, requirements common to several blocks need only be implemented once.

Different implementations can be explored and the trade-offs between speed and extra components can be investigated. Our approach to design is integrated with formal proof. We can safely explore design alternatives which would otherwise require manual rewriting of the specification. In our example we have presented two solutions which result from different unfoldings of the recursive definition.

Formal logic can treat abstractions such as MICRO on the same level as a component. Implementation level decisions (such as clock speed and latency) can be calculated during the course of the design.

Rearranging equations may appear trivial, since the designer instinctively *knows* that addition and multiplication are associative. In the case of finite precision arithmetic, however, addition is not necessarily associative and such a rearrangement is a common source of error. Our methodology ensures correctness by allowing the reassociation while keeping track of the conditions under which it is valid (such as no overflows occurring).

5.5.2   The Real World

There is currently little machine assistance for the passage from functional specification to structural realization. This has led to many expensive bugs in both custom and volume parts. We have outlined a formal basis for tools which support design refinement. The major benefit is to have machine assistance at an earlier stage of the design process. An interactive design assistant based on the representation of behaviour will give high-level feedback to the system designer and will expose logical errors and incorrect assumptions at this early stage. We have outlined a formal basis for system design. However, we have elided many factors which will be important in practical application.

Foremost among these is the difficulty of formalising the specification; it is almost as difficult to be confident that the specification expresses our needs as it is to be confident of the correctness of an implementation. Animation of the specification is commonly used to gain confidence, but many high-level specifications will be nondeterministic to allow flexibility in the choice of implementation strategy. In principle, it is possible to derive consequences of the specification formally. However, there is little experience of this approach in practical situations.

To apply these methods in industrial practice, considerable effort and expense is required to generate and maintain libraries of system components. Formal representation of the behavioural aspects of these libraries will provide greater reusability and better maintainability, but these benefits are deferred; the costs are immediate. There is also a problem of education; today's engineer is not accustomed to formal thought, and the typical formalist is not capable of engineering judgement.

5.6 Acknowledgements

In compiling these notes, I am indebted to my co-authors for material, adopted and, sometimes, adapted, from previous works, particularly Simon Finn, Mick Francis and Eleanor Mayger. They and others at Abstract Hardware Limited have engineered, in the LAMBDA / DIALOG system, an implementation that supports formal system design and thus the development of this approach. The bibliography acknowledges many other sources, I have also benefited from discussions with many researchers in hardware verification; Mike Gordon, in particular, introduced me (and many others) to the application of computer-assisted formal reasoning to hardware verification.

5.7 References

[1] Abstract Hardware Limited, Brunel University, Howell Building, Uxbridge, Middlesex UB8 3PH. *Lambda/Dialogue Version 3.0 Documentation*, 1990.

[2] H. Barrow. Proving the correctness of hardware designs. *VLSI Design*, pages 64–77, July 1984.

[3] G.C. Bell and A. Newell. *Computer Structures: Reading and Examples*. McGraw-Hill, 1971.

[4] R.E. Bryant. Symbolic verification of MOS circuits. In *1985 Chapel Hill Conference on VLSI*, pages 419–438, 1985.

[5] R.E. Bryant. Graph-based algorithms for Boolean function manipulation. *IEEE Transactions on computers*, C-35(8):677–691, August 1986.

[6] Herbert B. Enderton. *A Mathematical Introduction to Logic*. Academic Press, 1972.

[7] M.P. Fourman. The logic of topoi. In Barwise, editor, *Handbook of Mathematical Logic*, pages 1053–1090. North-Holland, 1977.

[8] M.P. Fourman. Verification using higher-order specifications. In *Proceedings of the Silicon Design Conference*, Wembley, 1986.

[9] M.P. Fourman, W.J. Palmer, and R.M. Zimmer. Using higher-order functions to describe hardware. In *Proceedings of the Electronic Design Automation Conference*, Wembley, 1987.

[10] M.P. Fourman, W.J. Palmer, and R.M. Zimmer. Proof and synthesis. In *Proceedings ICCD '88*, Rye Brook, NY, 1988.

[11] Michael J. Gordon, Robin Milner, and Christopher P. Wadsworth. A mechanised logic of computation. In *Edinburgh LCF: Lecture Notes in Computer Science—Volume 78*. Springer-Verlag, 1979.

[12] Mike Gordon. HOL: A machine oriented formulation of higher-order logic. Technical Report 68, University of Cambridge, 1985.

[13] Mike Gordon. Why higher-order logic is a good formalism for specifying and verifying hardware. In G. Milne and P.A. Subrahmanyam, editors, *Formal Aspects of VLSI Design*. North-Holland, 1986.

[14] Mike Gordon. HOL: A proof generating system for higher-order logic. In G. Birtwhistle and P.A. Subrahmanyam, editors, *VLSI Specification, Verification and Synthesis*. Kluwer Academic Publishers, 1988.

[15] F.K. Hanna and N. Daeche. Specification and verification using higher-order logic: A case study. In G.J. Milne and P.A. Subrahmanyam, editors, *Formal Aspects of VLSI Design*. North-Holland, 1986.

[16] F.K. Hanna, M. Longley, and N. Daeche. Formal synthesis of digital systems. In Dr. Luc Claesen, editor, *IMEC-IFIP International Workshop on Applied Formal Methods for Correct VLSI Design, Volume 2*, pages 532–548. Elsevier Science Publishers, B.V. North-Holland, Amsterdam, 1989.

[17] C.A.R. Hoare. *Communicating Sequential Processes*. Prentice-Hall International Series in Computer Science, 1985.

[18] W.A. Hunt. *FM8501: A Verified Microprocessor*. PhD thesis, Institute of Computer Science, The University of Texas at Austin, 1986.

[19] W.A. Hunt. A verified microprocessor. In *Proceedings of the IFIP Working Group 10.2 International Working Conference, HDL Descriptions to Guaranteed Correct Circuit Designs*, pages 85–115, Grenoble, France, September 9–11 1986.

[20] W.A. Hunt. The mechanical verification of a microprocessor design. In *HDL Descriptions to Guaranteed Correct Circuit Designs*, pages 89–132. North-Holland, 1987.

[21] Warren A. Hunt. Microprocessor design verification. *Journal of Automated Reasoning*, 1989. (To appear).

[22] J.J. Joyce, G. Birtwhistle, and M. Gordon. Verification and implementation of a microprocessor. In G. Birtwhistle and P.A. Subrahmanyam, editors, *VLSI Specification, Verification and Synthesis*. Kluwer Academic Publishers, 1988.

[23] C.E. Leiserson and J.B. Saxe. Retiming synchronous circuitry. Technical Report 13, Digital Systems Research Center, Palo Alto, California 94301, 1986.

[24] Carver Mead and Lynn Conway. *Introduction to VLSI Systems*. Addison-Wesley Publishing Company, 1980.

[25] T.F. Melham. Using recursive types of reason about hardware in high order logic. In *IFIP WG 10.2 International Working Conference on the Fusion of Hardware Design and Verification*, pages 26–49. North-Holland, 1988.

[26] Thomas F. Melham. Abstraction mechanisms for hardware verification. In Birtwhistle and Subrahmanyam, editors, *VLSI Specification, Verification, and Synthesis*, pages 267–291. Kluwer Academic Publishers, 1988.

[27] R. Milner. *A Calculus of Communicating Systems*. Springer-Verlag, LNCS 92, 1980.

[28] R. Milner. How ML evolved. *Polymorphism (the ML/LCF/Hope Newsletter*, 1(1), January 1983.

[29] R. Milner. *Communication and Concurrency*. Prentice Hall, 1989.

[30] R. Milner, M. Tofte, and R. Harper. *The Definition of ML*. The MIT Press, Cambridge, Massachusetts, 1990.

[31] Ben Moskowski. A temporal logic for multilevelreasoning about hardware. *IEEE Computer*, 2(18):10–19, February 1985.

[32] L. Paulson. Natural deduction proof as higher-order resolution. *Logic Programming*, 3:237–258, 1987.

[33] Daniel Weise. Constraints, abstraction and verification. In M. Leeser and G. Brown, editors, *MSI Workshop on Hardware Specification, Verification and synthesis: Mathematical Aspects*. Cornell University, Ithaca, New York, Springer Verlag, New York, 1989.

Formal Methods for VLSI Design
J. Staunstrup (Editor)
Elsevier Science Publishers B.V. (North-Holland)
© IFIP, 1990

Chapter 6

Synthesis of Asynchronous VLSI Circuits

Alain J. Martin[1]

6.1 Introduction

We present a synthesis method for the design of delay-insensitive VLSI circuits, i.e., VLSI circuits whose correct operation is independent of the delays in operators and wires, with the exception of so-called "isochronic forks."

A circuit is first designed as a concurrent program in a communicating-processes notation inspired by CSP. It is then compiled into a network of digital gates by a series of semantics-preserving program transformations. The network of gates is transformed into a layout for the chosen technology—usually CMOS—by an automatic placement and routing program. Hence, the circuits obtained are correct by construction.

The method has been applied both with "hand compilation" and automatic compilation to a series of non-trivial problems such as distributed mutual exclusion, fair arbitration, routing automata, FIFOs, stacks, and the design of a general purpose, 16-bit microprocessor. All fabricated chips have been found correct on "first silicon." They are also remarkably efficient: In 1.6 micron SCMOS, the microprocessor runs at 18 MIPS.

We shall first introduce the "source code," and discuss the design of hardware algorithms as communicating processes. Secondly, we shall describe the

---

[1]Department of Computer Science, California Institute of Technology, Pasadena CA 91125, USA. e-mail: alain@vlsi.caltech.edu

"object code," called production rules, and we shall relate it to the MOS technology.

We shall then present the different steps of the transformation and illustrate them with examples taken from the design of the microprocessor.

6.2   Communicating Hardware Processes

The source notation is a program notation and not a hardware description language. It is inspired by C.A.R. Hoare's CSP[5] and E.W. Dijkstra's guarded commands[4], and is based on assignment and process communication by message-passing.

6.2.1   Data Types and Assignment

The only basic data type is the boolean. The other types—integer and floating point—are collections of booleans that we represent in the PASCAL record notation.

For b boolean, the command $b := \textbf{true}$, also denoted $b\uparrow$, is the assignment of the value **true** to b. Similarly, the command $b := \textbf{false}$, also denoted $b\downarrow$, is the assignment of the value **false** to b.

An integer of length n is a predefined record type consisting of n boolean components ("fields" in the PASCAL jargon). For instance, if x is declared as the integer 8, then x is a collection of the 8 boolean variables: $x.0$, $x.1$, $x.2,\ldots, x.7$.

The existence of this predefined record type for integers does not preclude the programmer from introducing other records to structure the data. For instance, in the program of the microprocessor, which we will introduce later, the integer variable i represents (contains) the currently executed instruction. This is declared as a record of several types depending on the type of the instruction. For ALU instructions and ordinary memory instructions, the type is:

$$
\begin{aligned}
alu \;=\; &\textbf{record}\\
&op : alu.15..alu.12\\
&x : alu.11..alu.8\\
&y : alu.7..alu.4\\
&z : alu.3..alu.0\\
&\textbf{end},
\end{aligned}
$$

where the field op contains the "opcode" of the instruction, the fields x and y contain the indices of the registers to be used as parameters of the instruction, and z contains the index of the register in which the result of the instruction execution is to be stored.

Since operations on boolean variables are the only primitive operations, any operation on other data types appearing in a program must be understood to be a shorthand notation or function call for the sequence of operations on boolean variables that will implement it.

For instance, given two integers x and y of the same length n, the assignment

$$y := x$$

is a shorthand notation for the multiple assignment

$$y.0, \ y.1, \ \ldots, \ y.(n-1) := x.0, \ x.1, \ \ldots, \ x.(n-1).$$

The multiple assignment of n expressions to n variables is different from the concurrent composition (which we will introduce shortly) of the n elementary assignments. In the multiple assignment, the n expressions are all evaluated before the results are assigned to the corresponding variables.

For the sake of clarity, we will use the usual integer arithmetic operators (for instance, $y := x + 1$ in the program of the microprocessor) in the first description of an algorithm. However, since these operators are not primitive constructs of the language, they are subsequently replaced with calls to functions that implement the operators in terms of boolean operations.

6.2.2 Arrays

The array mechanism is used as an address-calculation mechanism, and is used when the identity of the element in a set of variables that is to be used for some action will be determined during the computation. For example, the processor uses three arrays: the instruction memory array, *imem*, whose index is the program counter, *pc*; the data memory, *dmem*; and the array of general-purpose registers, *reg*. Hence, the execution of a *load* instruction, i, is described by the assignment:

$$reg[i.z] := dmem[reg[i.x] + reg[i.y]].$$

6.2.3 Composition Operators

There are three composition operators (also called "constructors"): the sequential operator, represented by the semicolon; the concurrent, or parallel, operator represented by the parallel bar, $\|$; and the coincident operator, represented by the bullet.

The semantics of the sequential composition $S1; S2$ are well known; let us only recall that the semicolon is associative, but of course not commutative.

We will assume that the semantics of the parallel composition are well known as well, although we are all aware of how difficult it is to define the semantics formally and simply.

We postulate that parallel composition is *weakly fair: If at a certain point of the computation of S1 $\|$ S2, x is the next atomic action of S1, then x will be executed after a finite number of atomic actions of S2.*

Parallel composition is associative and commutative.

The bullet operator is used solely to compose communication commands. (Communication commands will be introduced later.) Furthermore, the co-incident composition of two communication commands is defined only if the two commands are *non-interfering: Two programs are non-interfering if a variable modified by one program is not used by the other program.*

For $S1$ and $S2$ non-interfering communication commands, if the executions of both $S1$ and $S2$ in a certain state of the computation terminate, then the execution of $S1 \bullet S2$ in that state terminates. Furthermore, the completion of $S1$ coincides with the completion of $S2$; i.e., $S1$ and $S2$ are completed in the same state of the computation. (We will return to this definition later when we define the notion of completion of a non-atomic action.)

The bullet operator is associative and commutative.

The bullet has the highest priority, followed by the semicolon, followed by the parallel bar:

$$S0 \bullet S1; S2 \parallel S3 \equiv ((S0 \bullet S1); S2) \parallel S3.$$

6.2.4   Control Structures

The two control structures are the *selection* and the *repetition* of Dijkstra's guarded commands. However, the VLSI programmer and the software pro-grammer adopt opposite attitudes towards non-determinism. Whereas the latter is encouraged to maximize non-determinism as a way to avoid unneces-sary choices, the former is requested to minimize non-determinism to reduce the high cost of arbitration in a direct VLSI implementation of a set of guarded commands.

It is very difficult, if at all possible, to determine at "compile-time" which selections require arbitration. We therefore introduce two sets of control struc-tures, a deterministic set and a non-deterministic set, and let the programmer explicitly indicate where arbitration is needed.

Selection

The execution of the deterministic selection command

$$[G_1 \rightarrow S_1 [\!] \ldots [\!] G_n \rightarrow S_n],$$

where G_1 through G_n are boolean expressions, S_1 through S_n are program parts, (G_i is called a "guard," and $G_i \rightarrow S_i$ is called a "guarded command") amounts to the execution of the arbitrary S_i for which G_i holds. At any time at most one guard holds. If none of the guards is **true** , the execution of the command is suspended until one guard is **true** .

The non-deterministic selection command

$$[G_1 \rightarrow S_1 | \ldots | G_n \rightarrow S_n]$$

is identical to the previous one, except that several guards may be **true** at the same time. In such a case, an arbitrary true guard is selected.

Repetition

The execution of the deterministic repetition command

$$*[G_1 \rightarrow S_1 [\!| \ldots [\!| G_n \rightarrow S_n],$$

where G_1 through G_n are boolean expressions, and S_1 through S_n are program parts, amounts to repeatedly selecting the arbitrary S_i for which G_i holds, and executing S_i. At any time, at most one guard holds. If none of the guards is **true** , the repetition terminates.

The non-deterministic repetition command

$$*[G_1 \rightarrow S_1 | \ldots | G_n \rightarrow S_n]$$

is identical to the previous one, except that several guards may be **true** at the same time. In such a case, an arbitrary true guard is selected.

$[G]$, where G is a boolean expression, stands for $[G \rightarrow \textbf{skip}]$, and thus for "wait until G holds." (Hence, "$[G]$; S" and $[G \rightarrow S]$ are equivalent.)

$*[S]$ stands for $*[\textbf{true} \rightarrow S]$ and, thus, for "repeat S forever."

Reactive Process Structure

From the preceding definitions, the operational description of the statement

$$*[[G_1 \rightarrow S_1 [\!| \ldots [\!| G_n \rightarrow S_n]]$$

is "repeat forever: Wait until some G_i holds; execute an S_i for which G_i holds." This structure, which we call "reactive," is used very frequently. For instance, the *server* processes in the distributed mutual exclusion example are reactive processes.

6.2.5 The Replication Construct

Both because of the restriction of basic operations to booleans and because of the high degree of concurrency of VLSI algorithms, such algorithms are characterized by an extensive use of *replication*. A typical example is that some action has to be performed (sequentially or concurrently) on all the boolean variables that represent an integer. Another example is that of an n-place buffer constructed as the concurrent composition of n identical one-place buffers.

The notation therefore contains a syntactic operator, called the replication construct, which makes it possible to "clone" any program part into a number of instances.

The replication mechanism is used to represent a fixed, finite, and non-empty list of syntactic objects. Operationally, we can say that the replication mechanism is used to generate a list of objects at compile-time. An element of the list is any program part. The concatenation operator of the list is any constructor or separator of the language. The constructors are the semicolon for sequential composition, and the comma and the parallel bar for parallel composition. The separators are the bar for guarded commands, and the blank and the comma for lists of declarations.

Recursion is the basic mechanism for creating such a list. Since it is often convenient to "unroll" the simplest form of tail recursion as an iteration mechanism, both iteration and recursion are available.

The construct for replication by iteration is defined as follows: If

- **op** is any constructor or separator,

- i is an integer variable, called the *running index*,

- the *range*, defined by $n..m$, where n and m are integer constants, is not empty, i.e., $n \leq m$,

- $S(i)$ is any program part in which i appears free,

then,

$$\langle \mathbf{op}\, i : n..m : S(i) \rangle \overset{\text{def}}{=} \begin{cases} S(n), & \text{if } n = m \\ S(n) \ \mathbf{op} \ \langle \mathbf{op} \ i : n+1..m : S(i) \rangle, & \text{if } n < m \end{cases}$$

For $n < m$, the definition is ambiguous if **op** is not associative. In this case the definition is taken to be equivalent to

$$S(n) \ \mathbf{op} \ (\langle \ \mathbf{op} \ i : n+1..m : S(i) \rangle).$$

The bracket notation for replication is borrowed from Chandy and Misra[3], who use it for defining so-called quantified expressions. Observe that a replication command is not a quantified expression.

For example, the construct $[\langle \| i : 0..3 : G(i) \rightarrow S(i) \rangle]$ expands to

$$
\begin{aligned}
&[G(0) \rightarrow S(0) \\
&\| G(1) \rightarrow S(1) \\
&\| G(2) \rightarrow S(2) \\
&\| G(3) \rightarrow S(3) \\
&].
\end{aligned}
$$

The construct

$$\langle ; i : 0..2 : x.i := y.((i+1)\textbf{mod3}) \rangle$$

expands to

$$x.0 := y.1; \quad x.1 := y.2; \quad x.2 := y.0.$$

Replication constructs can be nested as in the following example:

$$\langle , i : 0..9 : \langle , j : 0..i : x(i,j) = 0) \rangle \rangle .$$

6.2.6 Procedures and Functions

Procedures are used with a simple parameter mechanism: A parameter is either *input* or *output*. For procedure p, declared as

$$\textbf{procedure } p(x : input; y : output); S$$

the call $p(a, b)$ is equivalent to the program part

$$x := a; S; b := y.$$

A parameter of a function is always an input parameter. For function y, declared as

$$\textbf{function } y(x); S$$

where S is the same program part as in procedure p, a statement Q containing the function call $y(a)$ is equivalent to the program part

$$p(a, b); Q_b^{f(a)},$$

where b is a "fresh" variable.

Tail recursion is allowed but not general recursion, since general recursion requires the construction, at execution time, of a stack whose size may vary with the parameters of the computation.

6.2.7   Concurrent Processes

The main building block for the construction of concurrent computations is
the *process*. In the design of the microprocessor for instance, each stage of
the pipeline is a process. Concurrent composition of processes is also the
main source of concurrency, although we allow the concurrent composition of
statements inside processes. In strict communicating-process design style, a
variable is local to a process, and communication among processes is uniquely
by way of message exchanges. In the design of the processor, we have violated this rule and allowed processes to share variables in a restricted way: A
variable of one process may be inspected by another process. (Whether this
relaxation of the locality rule is a useful extension or a weakness of the flesh
is not clear at the moment. More experimentation is necessary.)

Hence, the most common structure for the body of a concurrent computation is the parallel construct:

$$p1 \parallel p2 \parallel \ldots \parallel pn$$

where $p1$ through pn are the names of processes that have been declared
beforehand. A process is used very much as a procedure is used: It is first
declared in a declaration statement and then called by using its name in
a statement. Several instances of the same process type can be called by
assigning different names. But, unlike procedures, each (instance of a) process
can be called only once.

Communication Commands, Ports, and Channels

Processes communicate with each other by using communication commands
on *ports*. A port of a process is paired with a port of another process to form
a *channel*. For the time being, we will assume that a channel is shared by
exactly two processes; later, we will generalize the definition to more than two
processes.

A process is either *elementary* or *composite*. The ports of an elementary
process are *external*: Each is to be connected by a channel to a port of another
process to form a composite process. The external ports of a process are
declared in the heading of the process, like the parameters of a procedure:

$$p \equiv \textbf{process}(R, L)$$

(Later on, we will add some type information to the declaration.) A composite
process, p, is the parallel composition of several processes. The ports of
a component process that are connected by a channel to ports of another
component process are internal to p. The ports of the components that are
left unconnected (dangling) are the external ports of p. The internal ports
and the channels are defined by channel declaration in the process body.

We use two equivalent naming mechanisms for ports and channels. The first one gives local names to ports and pairs the two ports of a channel. For example, let two processes, $p1$ and $p2$, share a channel with port X in $p1$ and port Y in $p2$. The declaration is as follows:

$$p1 \equiv \textbf{process}(X)\ldots\textbf{end}$$
$$p2 \equiv \textbf{process}(Y)\ldots\textbf{end}$$
$$p1 \parallel p2$$
$$\textbf{chan}(p1.X, p2.Y)$$

The second mechanism gives global names to channels, and uses the channel names for all ports of the same channel. For instance, the same two processes would be described as:

$$p1 \equiv \textbf{process}(C)\ldots\textbf{end}$$
$$p2 \equiv \textbf{process}(C)\ldots\textbf{end}$$
$$p1 \parallel p2$$
$$\textbf{chan } C$$

We prefer local names for ports when the processes involved are identical (as in the case of the server processes in the distributed mutual-exclusion example); we prefer global names when the processes are different because this reduces the nomenclature. (We have used global names in the description of the processor.)

If the channel is used only for synchronization between the processes, the name of the port is sufficient for identifying a communication on this port. For instance, in the program for distributed mutual exclusion, the channel between a "master" process and its "server" process is identified with port D in the master and port U in the server, and is used for synchronization only.

Semantics of Synchronization

If two processes, $p1$ and $p2$, share a channel with port X in $p1$ and port Y in $p2$, then, at any time, the number of completed X-actions in $p1$ will equal the number of completed Y-actions in $p2$; in other words, the completion of the n-th X-action "coincides" with the completion of the n-th Y-action.

If, for example, $p1$ reaches the n-th X-action before $p2$ reaches the n-th Y-action, the completion of X is suspended until $p2$ reaches Y. The X-action is then said to be *pending*. When, thereafter, $p2$ reaches Y, both X and Y are completed. The predicate "X is pending" is denoted as qX.

If, for an arbitrary command A, cA denotes the number of completed A-actions, the semantics of a pair (X, Y) of communication commands is expressed by the two axioms:

$$cX = cY$$
$$\neg qX \quad \vee \quad \neg qY$$

Probe

Instead of the usual selection mechanism by which a set of pending communication actions can be selected for execution, we provide a general boolean command on channels, called the *probe*. The definition of the probe[6] states that in process $p1$, the probe command $\overline{X}$ has the same value as $\mathbf{q}Y$, and symmetrically in $p2$.

Hence, the guarded command $\overline{X} \rightarrow X$ guarantees that the X-action is not suspended.

Remark: In view of our declared intention to implement processes in a distributed and delay-insensitive way, our choice of definitions for communication may already puzzle some readers: The definition of $A1$ relies on the *simultaneous* completion of two actions in two different processes, and the value of the probe in one process is supposed to be *identical* to a suspended state of another process. A short explanation is that we have chosen definitions of completion and suspension that are unorthodox but valid!   ☐

Example

Process *sel* repeatedly performs communication action X or communication action Y, whichever can be completed; *sel* is blocked if and only if neither X nor Y can be completed. The program body of *sel* is:

$$*[[\overline{X} \rightarrow\ X [\overline{Y} \rightarrow Y]]$$

Obviously, process *sel* is not fair, because of the non-deterministic choice of a guard when both guards are true. Negated probes make it possible to transform *sel* into a fair version, *fsel*, whose body is:

$$*[\,[\overline{X} \rightarrow X;\ [\overline{Y} \rightarrow Y [\neg\overline{Y} \rightarrow \mathbf{skip}]$$
$$[\overline{Y} \rightarrow Y;\ [\overline{X} \rightarrow X [\neg\overline{X} \rightarrow \mathbf{skip}]$$
$$]\,]\ .$$

This example illustrates the fact that negated probes are necessary for implementing fairness.

Communication

Matching communication actions are also used to implement a form of distributed assignment statement, to "pass messages" as it is often said. In that case, the pair of commands is specified to consist of an input command and an output command by adjoining to them the symbols "?" and "!", respectively. For example, $X?$ is an input command and then X is an input port, and $Y!$ is an output command, and then Y is an output port.

Communication axiom. *Let $X?u$ and $Y!v$ be matching, where u is a process variable, and v is an expression of the same type as u. The communication implements the assignment $u := v$. In other words, if $v = V$ before the communication, $u = V$ and $v = V$ after the communication.*

6.2.8 Examples

In this section, we illustrate the notation with a number of typical examples. The programs are given with a brief informal explanation. All proofs of correctnes are omitted.

Buffers

The following program constructs a one-place buffer. The process inputs 8-bit integer messages on the input port L, and outputs them in the same order on the output port R. (The process can be implemented without introducing the internal variable x.)

$$BUF1 \equiv \textbf{process}(L?int(8); R!int(8))$$
$$x : int(8)$$
$$*[L?x; R!x]$$
$$\textbf{end}$$

A buffer of size n can be constructed as the linear composition of n one-place buffers.

$$BUF(n) \equiv \textbf{process}(L?int(8), R!int(8))$$
$$p(i : 0..n-1) : BUF1$$
$$\langle \|i : 0..n-1 : p(i)\rangle$$
$$\textbf{chan}\langle i : 0..n-2 : (p(i).R, p(i+1).L)\rangle$$
$$p(0).L = BUF(n).L$$
$$p(n-1).R = BUF(n).R$$
$$\textbf{end}$$

Distributed Mutual Exclusion on a Ring of Processes

An arbitrary number (> 1) of cyclic automata, called "masters," make independent requests for exclusive access to a shared resource. The circuit should handle the requests from the masters in such a way that

1. Any request is eventually granted, and

2. there is at most one master using the shared resource at any time.

The masters are independent of each other: They do not communicate with each other, and the activity of a master not using the resource should not influence the activity of other masters.

A master, M, communicates with its private server, m. When M wants to use the shared resource (M is said to be a *candidate*), it issues a request to m. When the request is accepted, M uses that resource (for a finite period of time), and then informs m that the resource is free again.

The servers are connected in a ring. At any time, exactly one (arbitrary) server holds a "privilege." Only the "privileged server" may grant the resource to its master and thereby guarantee mutual exclusion on the access to the resource. A non-privileged server transmits a request from its master (or from its left-hand neighbor) to its right-hand neighbor. A request circulates to the right (clockwise) until it reaches a server whose master is a candidate (this server ignores the request until it has served its master) or reaches the privileged server. The privileged server reflects the privilege to the left (counter-clockwise) until it reaches the server that generated the request. This server then becomes privileged, and may grant the resource to its master. The strategy of passing requests clockwise and reflecting the privilege counterclockwise has two important advantages: First, no boolean message need actually be transmitted; second, no message need be reflected, as the completion of a pending request is interpreted as passing the privilege.

$$
\begin{aligned}
master \;&\equiv\; *[\ldots D; CS; D] \\
server \;&\equiv\; *[\,[\,\overline{U} \to [b \to skip \| \neg b \to R]; U; U; b\uparrow \\
&\qquad\; |\overline{L} \to [b \to skip \| \neg b \to R]; L; b\downarrow \\
&\qquad]].
\end{aligned}
$$

The boolean b is used to encode the presence of the privilege. The non-deterministic bar indicates that both guards may be true at the same time, and therefore arbitration has to take place. We can describe a system in which n servers are connected in a ring by first defining a process *pair* consisting of a master and a server, and then connecting n pairs in a ring:

$$
\begin{aligned}
pair \equiv\; &\textbf{process}(L, R) \\
&m : server \\
&M : master \\
&(m\|M) \\
&\textbf{chan}(m.U, M.D) \\
&\textbf{end} \\
ring \equiv\; &\textbf{process} \\
&p(i : 0..n-1) : pair \\
&\langle\,\|i : 0..n-1 : p(i)\rangle \\
&\textbf{chan}\langle i : 0..n-1 : (p(i).R, p((i+1)\,\textbf{mod}\,n).L)\rangle \\
&\textbf{end}
\end{aligned}
$$

(For a complete description and proof of correctness, see [7].)

An Asynchronous Microprocessor

The processor is first described as a sequential program, which is then transformed into a set of concurrent processes so as to increase the concurrency in the execution of a sequence of instructions by pipelining. The sequential program is a non-terminating loop, each step of which is a *FETCH* phase followed by an *EXECUTE* phase.

$$
\begin{aligned}
*[\, FETCH : \ & i, pc := imem[pc], pc + 1; \\
& [\mathit{off}(i) \rightarrow \mathit{offset}, pc := imem[pc], pc + 1; \\
& [\![\neg \mathit{off}(i) \rightarrow skip \\
&]; \\
EXECUTE : \ & [alu(i) \rightarrow (reg[i.z], f) := \\
& \qquad\qquad aluf(reg[i.x], reg[i.y], i.op, f) \\
& [\![ld(i) \rightarrow reg[i.z] := dmem[reg[i.x] + reg[i.y]] \\
& [\![st(i) \rightarrow dmem[reg[i.x] + reg[i.y]] := reg[i.z] \\
& [\![ldx(i) \rightarrow reg[i.z] := dmem[\mathit{offset} + reg[i.y]] \\
& [\![stx(i) \rightarrow dmem[\mathit{offset} + reg[i.y]] := reg[i.z] \\
& [\![lda(i) \rightarrow reg[i.z] := \mathit{offset} + reg[i.y] \\
& [\![stpc(i) \rightarrow reg[i.z] := pc \\
& [\![jmp(i) \rightarrow pc := reg[i.y] \\
& [\![brch(i) \rightarrow [cond(f, i.cc) \rightarrow pc := pc + \mathit{offset} \\
& \qquad\qquad\quad [\![\neg cond(f, i.cc) \rightarrow skip \\
& \qquad\qquad\quad] \\
&] \\
]. &
\end{aligned}
$$

The variables of the program are the following: As we already mentioned, variable i contains the instruction currently being executed. All instructions contain an *op* field describing the *opcode*. The parameter fields depend on the types of the instructions. The most common ones, those for ALU, load, and store instructions, consist of the three parameters x, y, and z. Variable cc contains the condition code field of the branch instruction, and f contains the *flags* generated by the execution of an *alu* instruction.

The two memories are described as the arrays *imem* and *dmem*. The index to *imem* is the program counter variable *pc*. Variable *offset* contains the offset field that extends certain instructions to the following word. The general-purpose registers are described as the array $reg[0 \ldots 15]$. Register $reg[0]$ is special: It always contains the value zero.

The function evaluation $(z, f) := aluf(x, y, op, f)$ evaluates an *alu* instruction with the opcode, *op*; parameters x and y; and the current value of the flags, f. The result is an integer, z, and a new value of the flags, f. The function, *aluf*, is not described in the program. The boolean functions used in guards all determine certain properties of the current instruction i and are assumed to be self-explanatory.

First Decomposition into Concurrent Processes

The first step of the decomposition consists in replacing the previous program with the program:

$$*[FETCH; E1!i; E2] \| *[E1?i; EXECUTE; E2]$$

We leave it as an exercise to the readers to convince themselves that this decomposition does not introduce concurrency. The concurrent program is strictly equivalent to the sequential one.

Concurrent activity between the two processes will be introduced by moving $E2$ forward in the code of $EXECUTE$ so that the $n+1$rst step of $FETCH$ can start before the nth step of $EXECUTE$ is finished. This refinement, and the further decomposition of $EXECUTE$ into several processes is not discussed here. The resulting program can be found in [9].

The rest of the exercise will concentrate on the further decomposition of $FETCH$. The practical way to exploit concurrency in $FETCH$ is through the implementation of the multiple assignments. We introduce a process for the instruction memory which communicates the next instruction at address pc by a communication action on channel ID. Observe that variable pc is shared by the two processes. We get the following program:

$$
\begin{aligned}
IMEM \quad &\equiv \quad *[ID!imem[pc]] \\
FETCH \quad &\equiv \quad *[\,(ID?i \| y := pc + 1); pc := y; \\
&\qquad [\,off(i) \rightarrow (ID?\,offset \| y := pc + 1); pc := y \\
&\qquad \| \neg off(i) \rightarrow skip \\
&\qquad]; E1!i; E2 \\
&\qquad] \\
EXEC \quad &\equiv \quad *[E1?i; EXECUTE; E2]
\end{aligned}
$$

Next, we delegate the execution of the assignments $y := pc + 1; pc := y$ to a separate process as follows:

$$
\begin{aligned}
FETCH \quad &\equiv \quad *[\,PCI1; ID?i; PCI2; \\
&\qquad [\,off(i) \rightarrow PCI1; ID?\,offset; PCI2 \\
&\qquad \| \neg off(i) \rightarrow skip \\
&\qquad]; E1!i; E2 \\
&\qquad] \\
PCADD \quad &\equiv \quad *[PCI1; y := pc + 1; PCI2; pc := y]
\end{aligned}
$$

(The reader worrying about the cost of these extra communications has to realize that the two pairs of communications $E1$ and $E2$, and $PC1$ and $PC2$ are each implemented as the two halves of the same communication action.)

6.3 The Object Code, Production Rules

6.3.1 Definitions

Production Rule. *A production rule (PR) is a construct of the form $G \mapsto S$, where S is either a simple assignment or an unordered list "$s1$, $s2$, $s3$, $\ldots$" of simple assignments, and G is a boolean expression called the guard of the PR.*

Example:

$$x \wedge y \mapsto z \uparrow$$
$$\neg x \mapsto u \uparrow, v \downarrow$$

The semantics of a PR are defined only if the PR is *stable*:

Stability. *A PR $G \mapsto S$ is said to be stable in a given computation, if, at any point of the computation, G either is **false** or remains invariantly **true** until the completion of S.*

Stability is not guaranteed by the implementation. It has to be enforced by the compilation procedure.

Execution of a PR. *An execution of the stable PR $G \mapsto S$ is an unbounded sequence of firings. A firing of $G \mapsto S$ with G **true** amounts to the execution of S. A firing of $G \mapsto S$ with G **false** amounts to a **skip**.*

If S is a list of several simple assignments, the execution of S is the concurrent execution of all asssignments of the list.

Production Rule Set. *A PR set is the concurrent composition of all PRs of the set.*

The only composition operation on two PR sets is the set union.

Theorem. *The implementation of two concurrent processes is the set union of the two PR sets implementing the processes and of the PR sets implementing the channels between the processes, if any.*

The proof follows from the associativity of the concurrent composition operator. The other operations on the PRs of a set are those allowed by the following properties:

- Multiple occurrences of the same PR are equivalent to one as a consequence of the idempotence of the concurrent composition.

- The two rules $G \mapsto S1$ and $G \mapsto S2$ are equivalent to the single rule $G \mapsto S1, S2$.

- The two rules $G1 \mapsto S$ and $G2 \mapsto S$ are equivalent to the single rule $G1 \vee G2 \mapsto S$.

PRs are *complementary* when they are of the type $G1 \mapsto x \uparrow$ and $G2 \mapsto x \downarrow$. We require that complementary PRs be *non-interfering*.

Non-Interference. *Two complementary PRs are non-interfering when* $\neg G1 \vee \neg G2$ *holds invariantly.*

It can be proven that, under the stability of each PR and non-interference among complementary PRs, the concurrent execution of the PRs of a set is equivalent to the following sequential execution:

$$*[select \ a \ PR \ with \ a \ true \ guard; \ fire \ the \ PR]$$

where the selection is weakly fair (each PR is selected infinitely often). From now on, we ignore the firings of a PR with a **false** guard; a firing will mean a firing of a PR with a **true** guard.

Until we return to these issues, we shall assume that the stability and non-interference requirements are fulfilled.

6.3.2   Switching circuits

Consider the canonical (stable) PR

$$b \mapsto z \downarrow$$

where b is a boolean expression in terms of a set of variables. These variables are used as gates of transistors implementing a switching circuit s corresponding to b: s is a series-parallel switching circuit between the ground node (also called GND) and z. GND has the constant value **false** . The other constant node, the power-node VDD, has the constant value **true** .

The switches are n-transistors whose gates are the variables of b, possibly negated. Furthermore, we have:

$$b \equiv \text{``there is a path from ground to } z \text{ in } s\text{''}$$

By construction of s, if b holds and remains stable, z is eventually set to **false** . (For this reason, s is called a *pull-down circuit*.) Hence, s is exactly the implementation of the production rule $b \mapsto z \downarrow$.

Using a symmetrical argument, we can show that the same series-parallel circuit as s, but with VDD and z connected, and whose switches are p-transistors, implements the production-rule:

$$bneg \mapsto z \uparrow ,$$

where *bneg* is derived from b by negating all variables. (This circuit is called a *pull-up circuit*.)

6.3.3 Operators

The two PRs that set and reset the same variable, like

$$b1 \mapsto z \uparrow$$
$$b2 \mapsto z \downarrow ,$$

are implemented as one operator.

Let $s1$ be the pull-up circuit corresponding to $b1$, and let $s2$ be the pull-down circuit corresponding to $b2$. The two circuits are connected through the common node z. Since non-interference has been enforced, $\neg b1 \vee \neg b2$ holds at any time. This guarantees the absence of a conducting path between power and ground when the operator is not firing. (A path may exist for a short time when the operator is firing.)

Definition. *The operator implementing the two rules is called "combinational" if $b1 \vee b2$ holds at any time, and "state-holding" otherwise.*

By definition, if the operator is combinational, there is always a conducting path between either VDD or GND and the output z. Hence, the value of the output is always a strong **false** value or a strong **true** value, and therefore the circuit corresponding to the composition of $s1$ and $s2$ is a valid implementation of the operator.

For example, PRs (1) and (2) together implement an inverter. The circuit of Figure 1 implements the *nand*-operator defined by the PRs

$$a \wedge b \mapsto z \downarrow$$
$$\neg a \vee \neg b \mapsto z \uparrow$$

If (5) is a state-holding operator, $\neg b1 \wedge \neg b2$ may hold in a certain state. In such a state, node z is isolated; there is no path between z and either VDD or GND. In MOS technology, an isolated node does not retain its value forever; eventually the charges leak away through the substrate and also through the transistors of the pull-up and pull-down circuits. If the PRs of the operator are fired frequently enough to prevent leakage, the implementation of Figure 6.1 can be used for a state-holding operator. Such an implementation is called *dynamic*.

Otherwise, it is necessary to add a storage element to the output node of a state-holding operator. Such an implementation is called *static*. In the sequel, we assume that only static implementations are used for state-holding operators.

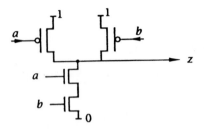

Figure 6.1: CMOS implementation of a NAND-operator

A standard CMOS implementation of such a storage element consists of two cross-coupled inverters (see Figure 6.2). This implementation inverts the value of z.

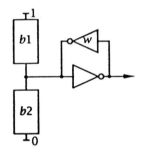

Figure 6.2: A static implementation of a state-holding operator

The "weak" inverter, marked with a letter w on the figure, connects z to either *VDD* or *GND* through a high resistance, so as to maintain z at its intended voltage value [18].

The implementation of a static state-holding operator is slightly more costly than that of a combinational operator because of the need for a storage device. Hence, given a pair of PRs that are not combinational, we may first try to modify the guards—under the invariance of the semantics—so as to make them combinational.

6.3.4   The Standard Operators

All operators of one or two inputs are used, and are therefore viewed as the standard operators.

One-Input Operators

The two operators with one input and one output are the *wire*:

$$x \, \underline{w} \, y \equiv \quad x \mapsto y \uparrow$$
$$\neg x \mapsto y \downarrow \, ,$$

and the *inverter*:

$$\neg x \, \underline{w} \, y \equiv \neg x \mapsto y \uparrow$$
$$x \mapsto y \downarrow \, .$$

Most operators we use have more inputs than outputs. But, in general, the components we design have as many outputs as inputs. Hence, we need to reset the balance by introducing at least one operator, the *fork*, with more outputs than inputs. A fork with two outputs is defined as:

$$x \, \underline{f} \, (y, z) \equiv \quad x \mapsto y \uparrow, z \uparrow$$
$$\neg x \mapsto y \downarrow, z \downarrow \, .$$

The wire and the fork are the only two operators that are not implemented as a pull-up/pull-down circuit—called a *restoring* circuit—but as a simple conducting interconnection between input and outputs.

The Wire as a Renaming Operator

Because the implementation of a wire is the same as that of a node, the wire behaves as a renaming operator when composed with another operator: The composition of an arbitrary operator O with output variable x with the wire $x \, \underline{w} \, y$ is equivalent to O in which x is renamed y. The composition of operator O with input variable x with the wire $y \, \underline{w} \, x$ is equivalent to O in which x is renamed y. (Observe that O can even be a wire.)

Unfortunately, the fork is not a renaming operator since the concurrent assignments to the different outputs of the fork are not completed simultaneously. In order to use a fork as a renaming operator, we will later have to make the timing assumption that such a fork is *isochronic*.

Combinational Operators with Two Inputs

We construct all functions B of two variables x and y such that

$$B \mapsto z \uparrow$$
$$\neg B \mapsto z \downarrow .$$

We get for B: $x \wedge y$, $x \vee y$, and $x = y$. We will not list the functions obtained by inverting inputs of B. (On the figures, a negated input or output is represented by a small circle on the corresponding line.) This gives the following set. The *and*, with the infix notation $(x, y) \underline{\wedge} z$, is defined as:

$$x \wedge y \mapsto z \uparrow$$
$$\neg x \vee \neg y \mapsto z \downarrow .$$

The *or*, with the infix notation $(x, y) \underline{\vee} z$, is defined as:

$$x \vee y \mapsto z \uparrow$$
$$\neg x \wedge \neg y \mapsto z \downarrow .$$

The *equality*, with the infix notation $(x, y) \underline{eq} z$, is defined as:

$$x = y \mapsto z \uparrow$$
$$x \neq y \mapsto z \downarrow .$$

State-Holding Operators with Two Inputs

Next, we construct all different two-input-one-output operators of the form

$$b1 \mapsto z \uparrow$$
$$b2 \mapsto z \downarrow$$

such that $\neg b1 \vee \neg b2$ holds at any time, but $b1 \neq \neg b2$. We select for $b1$ either $x \wedge y$, or $x \vee y$, or $x = y$. For each choice of $b1$, we construct $b2$ as any of the effective strengthenings of $\neg b1$.

For $b1 \equiv (x \wedge y)$, we get for $b2$: $\neg x \wedge \neg y$, $\neg x \wedge y$, $\neg x$, and $x \neq y$. The first three choices of $b2$ lead to the following state-holding operators:
The *C-element*
$$(x, y) \underline{C} z \equiv \quad x \wedge y \mapsto z \uparrow$$
$$\neg x \wedge \neg y \mapsto z \downarrow .$$

(The C-element was introduced by David Muller, and described in [16] .)
The *switch*
$$(x, y) \underline{sw} z \equiv \quad x \wedge y \mapsto z \uparrow$$
$$\neg x \wedge y \mapsto z \downarrow .$$

The *asymmetric C-element*

$$(x, y) \; \underline{aC} \; z \; \equiv \quad x \wedge y \; \mapsto z \uparrow$$
$$\neg x \; \mapsto z \downarrow \; .$$

For $b2 \equiv (x \neq y)$, we get the operator

$$x \wedge y \mapsto z \uparrow$$
$$x \neq y \mapsto z \downarrow \; .$$

But, if the stability condition is fulfilled, this operator is not state-holding. Because of the stability requirement, the state in which $\neg x \wedge \neg y$ holds—the "storage state"—can only be reached from states $x \wedge \neg y$ and $\neg x \wedge y$. In both states, $\neg z$ holds, and, therefore, $\neg z$ holds in the storage state. Hence, we can weaken the guard of the second PR as $(x \neq y) \vee (\neg x \wedge \neg y)$, i.e., $\neg x \vee \neg y$. Hence, the operator is equivalent to the *and*-operator $(x, y) \bigtriangleup z$.

For $b1 \equiv (x \vee y)$, no effective strengthening of $\neg b1$ is possible.

For $b1 \equiv (x = y)$, we get the operator:

$$x = y \mapsto z \uparrow$$
$$x \wedge \neg y \mapsto z \downarrow \; .$$

But if the stability condition is fulfilled, this operator is not state-holding for the same reasons that the operator with $b1 \equiv x \wedge y$ and $b2 \equiv (x \neq y)$ is not.

Flip-Flop

The canonical form we choose for the *flip-flop* is :

$$(x, y) \; \underline{ff} \; z \; \equiv \quad x \mapsto z \uparrow$$
$$\neg y \mapsto z \downarrow \; ,$$

which requires the invariance of $\neg x \vee y$ to satisfy non-interference. Observe that the flip-flop $(x, y) \underline{ff} z$ can always be replaced with the *C*-element $(x, y) \underline{C} z$ but not vice versa.

6.3.5 Multi-Input Operators

We use n-input *and, or, C-element*, whose definitions are straightforward. We use a *multi-input flip-flop* defined as:

$$(x_1, \ldots, x_k, y_1, \ldots, y_l) \; \underline{mff} \; z \; \equiv \quad \bigvee i : x_i \mapsto z \uparrow$$
$$\bigvee i : \neg y_i \mapsto z \downarrow$$

where $(\forall i : \neg x_i) \vee (\forall i : y_i)$.

We also use the combinational *if*-operator—sometimes called *multiplexer*—defined as:

$$(x, y, z) \underline{if} \, u \equiv \quad (x \wedge y) \vee (\neg x \wedge z) \mapsto u \uparrow$$
$$(x \wedge \neg y) \vee (\neg x \wedge \neg z) \mapsto u \downarrow \, .$$

The most general and most often used operator is the *generalized C-element*, of which all other forms of C-elements are a special case. It implements a pair of PRs

$$B1 \mapsto x \uparrow$$
$$B2 \mapsto x \downarrow$$

in which $B1$ and $B2$ are arbitrary conjunctions of elementary terms. (As usual, the two guards have to be mutually exclusive.) For example:

$$a \wedge b \wedge \neg c \mapsto x \uparrow$$
$$\neg a \wedge d \mapsto x \downarrow$$

can be directly implemented with a generalized C-element. Observe that the limiting factor for the size of the guards is not the number of inputs, but the number of terms in a conjunction.

6.3.6   Arbiter and Synchronizer

So far, we have considered only PR sets in which all guards are stable and non-interfering. But we shall have to implement sets of guarded commands—selections or repetitions—in which the guards are *not* mutually exclusive, as in the probe selection example. Therefore, we need at least one operator that provides a non-deterministic choice between two **true** guards. We will introduce the arbiter in an example.

6.4   The Compilation Method

6.4.1   Process Decomposition

The first step of the compilation, called *process decomposition*, consists in replacing one process with several processes by application of the following **Decomposition rule:** *A process, P, containing an arbitrary program part, S, is semantically equivalent to two processes, $P1$ and $P2$, where $P1$ is derived from P by replacing S with a communication action, C, on the newly introduced channel (C, D) between $P1$ and $P2$, and $P2$ is the process* $*[[\overline{D} \rightarrow S; D]]$.

The structure of $P2$ will be used so frequently that we introduce an operator to denote it: the *call* operator. We denote it by (D/S), and we say that D *calls* (or *activates*) S.

Observe that process decomposition does not introduce concurrency. Although $P1$ and $P2$ are potentially concurrent, they are never active concurrently; $P2$ is activated from $P1$, much as a procedure or a coroutine would be. The newly created subprocesses may share variables; but, since the subprocesses are never active concurrently, there is no conflicting access to the shared variables. The subprocesses may also share channels; this will require a special implementation for such channels. Decomposition is applied for each construct of the language. For construct S, the corresponding process $P2$ can be simplified as follows:

- If S is the selection $[B_1 \rightarrow S_1 \llbracket B_2 \rightarrow S_2]$, $P2$ is simplified as

$$*[[\overline{D} \wedge B_1 \rightarrow S_1; D$$
$$\llbracket \overline{D} \wedge B_2 \rightarrow S_2; D \qquad (6.1)$$
$$]]\,.$$

- If S is the repetition $*[B_1 \rightarrow S_1 \llbracket B_2 \rightarrow S_2]$, $P2$ is simplified as

$$*[[\overline{D} \wedge B_1 \rightarrow S_1$$
$$\llbracket \overline{D} \wedge B_2 \rightarrow S_2$$
$$\llbracket \overline{D} \wedge \neg B_1 \wedge \neg B_2 \rightarrow D \qquad (6.2)$$
$$]]\,.$$

- The assignment $x := B$, where B is an arbitrary boolean expression, is implemented as the selection $[B \rightarrow x \uparrow \llbracket \neg B \rightarrow x \downarrow]$, which gives for $P2$

$$*[[\overline{D} \wedge B \rightarrow x \uparrow; D$$
$$\llbracket \overline{D} \wedge \neg B \rightarrow x \downarrow; D \qquad (6.3)$$
$$]]\,.$$

The generalizations to the cases of an arbitrary number of guarded commands in selection and repetition are obvious. All assignments to the same variable are also grouped in the same process. Process decomposition is applied repeatedly until the right-hand side of each guarded command is a straight-line program.

Process decomposition makes it possible to reduce a process with an arbitrary control structure to a set of subprocesses of only two different types: either a (finite or infinite) sequence of communication actions, or a repetition of type 6.1, 6.2, or 6.3.

6.4.2   Handshaking Expansion

The next step of the transformation, the *handshaking expansion*, replaces each communication action in a program with its implementation in terms of elementary actions, and each channel with a pair of wire-operators. We shall first ignore the issue of message transmission and implement only the synchronization property of communication primitives.

Channel (X, Y) is implemented by the two wires $(xo \ \underline{w} \ yi)$ and $(yo \ \underline{w} \ xi)$. If X belongs to process $P1$ and Y to process $P2$, then xo and xi belong to $P1$, and yo and yi to $P2$. Initially, xo, xi, yo, and yi—which we will call the "handshaking variables of (X, Y)"—are **false**. Assume that the program has been proven to be deadlock-free and that we can identify a pair of matching actions X and Y in $P1$ and $P2$, respectively. We replace X and Y by the sequences U_x and U_y, respectively, with:

$$U_x \equiv xo \uparrow; \ [xi]$$
$$U_y \equiv [y_i]; \ yo \uparrow \ . \tag{6.4}$$

Also:

$$xo \mapsto yi \uparrow$$
$$\neg xo \mapsto yi \downarrow$$
$$yo \mapsto xi \uparrow \tag{6.5}$$
$$\neg yo \mapsto xi \downarrow,$$

by definition of the wires. By 6.4 and 6.5, any concurrent execution of $P1$ and $P2$ contains the sequence of assignments:

$$xo \uparrow; \ yi \uparrow; \ yo \uparrow; \ xi \uparrow \ .$$

Simultaneous Completion of Non-Atomic Actions

We introduce a definition of *completion* of a non-atomic action which makes it possible to use the notion of simultaneous completion of two non-atomic actions.

By definition, the execution of an atomic action is considered instantaneous, and thus the simultaneous completion of two atomic actions does not make sense. (Atomic actions are simple assignments $x \uparrow$ and $x \downarrow$, and evaluation of simple guards, i.e., guards containing one variable. A wait action of the form $[ai]$ is a non-atomic action that may be treated as the repetition $*[\neg ai \rightarrow skip]$.)

A non-atomic action is *initiated* when its first atomic action is executed. A non-atomic action is *terminated* when its last atomic action is executed.

For non-atomic actions, the notion of completion does not coincide with that of termination. A non-atomic action might be considered completed even if it has not terminated, i.e., even if some atomic actions that are part of the action have not been executed. The definition of suspension is derived from that of completion.

Definition. *A non-atomic action X is completed when it is initiated and it is guaranteed to terminate, i.e., when all possible continuations of the computation contain the complete sequence of atomic actions of X.*

The above definition can be further explained as follows: Consider a prefix
$t1$ of an arbitrary *trace* of a computation. (A trace is a sequence of atomic
actions corresponding to a possible execution of the program.) The completion of X is identified with the point in the computation where $t1$ has been
completed, if 1) X is initiated in $t1$, and 2) all possible sequences $t2$, such that
$t1$ extended with $t2$ is a valid trace of the computation, contain the remaining atomic actions of X. *Hence, the completions of two non-atomic actions
coincide if their completion points coincide.*

(Observe that there may be several points in a trace that can act as completion point, which makes it easier to align the two completion points of two
overlapping sequences so as to implement the bullet operator.)

Definition. *Between initiation and completion, an action is suspended.*

These definitions of completion and suspension are valid because they
satisfy the three semantic properties of completion and suspension that are
used in correctness arguments, namely:

- $\{cX = x\}\ X\ \{cX = x + 1\}$,

- $qX \Rightarrow pre(X)$, where $pre(X)$ is any precondition of X in terms of the
  program variables and auxiliary program variables,

- If X is completed, eventually X is terminated.

These definitions will be used to implement the bullet operator and the
communication primitives as defined by axioms $A1$ and $A2$. Consider the
interleaving of U_x and U_y. At the first semicolon, i.e., after $xo\uparrow$, U_x has been
initiated, but cannot be considered completed since the valid continuation
that does not contain U_y does not contain the rest of U_x. At the second
semicolon, both U_x and U_y have been initiated, and thus, all continuations
contain the rest of the interleaving of U_x and U_y. Hence, U_x and U_y are
guaranteed to terminate when they are both initiated, i.e., they fulfil $A1$ and
$A2$.

Four-phase Handshaking

Unfortunately, when the communication implemented by U_x and U_y terminates, all handshaking variables are **true**. Hence, we cannot implement the
next communication on channel (X, Y) with U_x and U_y. However, the complementary implementation can be used for the next matching pair, namely:

$$D_x \equiv xo\downarrow;\ [\neg xi]$$
$$D_y \equiv [\neg yi];\ yo\downarrow\ .$$

The solution consisting in alternating U_x and D_x as an implementation of X,
and U_y and D_y as an implementation of Y, is called *two-phase handshaking*,

or *two-cycle signaling*. Since it is in most cases impossible to determine syntactically which X- or Y-actions follow each other in an execution, the general
two-phase handshaking implementations require testing the current value of
the variables as follows:

$$xo := \neg xo; \ [xi = xo]$$
$$[yi \neq yo]; \ yo := \neg yo \ .$$

In general, we prefer to use a simpler solution, known as *four-phase handshaking*, or *four-cycle signaling*. In a four-phase handshaking protocol, X-actions
are implemented as "$U_x; D_x$" and Y-actions as "$U_y; D_y$". Observe that the
D-parts in X and Y introduce an extra communication between the two processes whose only purpose is to reset all variables to **false**.

Both protocols have the property that for a matching pair (X, Y) of actions, the implementation is not symmetrical in X and Y. One action is
called *active* and the other one *passive*. The four-phase implementation, with
X active and Y passive, is:

$$X \equiv xo\uparrow; \ [xi]; \ xo\downarrow; \ [\neg xi] \tag{6.6}$$

$$Y \equiv [yi]; \ yo\uparrow; \ [\neg yi]; \ yo\downarrow \ . \tag{6.7}$$

(We will introduce an alternative form of active implementation, called
lazy active.) Although four-phase handshaking contains twice as many actions as two-phase handshaking, the actions involved are simpler and are more
amenable to the algebraic manipulations we shall introduce later. When operator delays dominate the communication costs, which is the case for communication inside a chip, four-phase handshaking will, in general, lead to
more efficient solutions. When transmission delays dominate the communication costs, which is the case for communication between chips, two-phase is
preferred.

Probe

A simple implementation of the probe $\overline{X}$ is xi, with X implemented as passive.
(Given our definition of suspension, the proof that this implementation of the
probe fulfills its definition is straightforward.)

A probed communication action $\overline{X} \rightarrow \ldots X$ is then implemented as

$$xi \rightarrow \ldots xo\uparrow; \ [\neg xi]; \ xo\downarrow \ .$$

Choice of Active or Passive Implementation

When no action of a matching pair is probed, the choice of which action should
be active and which passive is arbitrary, but a choice has to be made. The

choice can be important for the composition of identical circuits. A simple rule is that, for a given channel (X, Y), all actions on one port (called the *active port*) are active, and all actions on the other port (called the *passive port*) are passive. If $\overline{X}$ is used, all X-actions are passive—with the obvious restriction that $\overline{Y}$ cannot be used in the same program.

However, we shall see that this criterion for choosing active and passive ports may conflict with another criterion related to the implementation of input and output commands.

Reshuffling

In 6.6 and 6.7, D_x and D_y are used only to reset all variables to **false**. Hence, provided that the cyclic order of the actions of 6.6 and 6.7 is maintained, the sequences D_x and D_y can be inserted at any place in the program of each of the processes without invalidating the semantics of the communication involved. This transformation, called *reshuffling*, may introduce a deadlock.

Reshuffling, which is the source of significant optimizations, will be used extensively. It is therefore important to know when it can be applied without introducing deadlock.

There are two simple cases where the reshuffling of sequence "$U_x; D_x; S$" into sequence "$U_x; S; D_x$" does not introduce deadlock:

- S contains no communication action, or
- X is an internal channel introduced by process decomposition.

6.4.3   Lazy-active protocol

Consider the active implementation of communication command X:

$$xo\uparrow;\ [xi];\ xo\downarrow;\ [\neg xi]\ .$$

We introduce an alternative protocol, called *lazy active*:

$$[\neg xi];\ xo\uparrow;\ [xi];\ xo\downarrow\ .$$

The lazy active protocol is derived from the active one by postponing wait action $[\neg xi]$ until the next communication on X, and by adding a vacuous wait action $[\neg xi]$ at the beginning of the first communication X. Hence, the lazy active protocol is a correct implementation when combined with a passive protocol.

The lazy active protocol is not identical to a passive protocol in which the input variable is replaced with its negation. In a passive protocol, the effective part (the upgoing part) of the protocol is $[xi];\ xo\uparrow$. In a lazy active one, the effective part is $xo\uparrow;\ [xi]$.

6.4.4   Production-rule Expansion

Production-rule expansion is the transformation from a handshaking expansion to a set of PRs. It is the most crucial and most difficult step of the compilation since it requires enforcing sequencing by semantic means. It consists of three steps: state assignment, guard strengthening, symmetrization.

6.4.5   Sequencing and Stability

Consider the handshaking expansion

$$S \equiv *[[w_0]; \; t_0; \; [w_1]; \; t_1; \ldots; [w_{n-1}]; \; t_{n-1}] \; .$$

The wait-conditions are boolean expressions, possibly identical to **true** , and the t_i are simple assignments. The extension to the case of multiple assignments between the wait-conditions is straightforward.

The next step of the compilation procedure—the *production rule expansion*— is the transformation of S into a semantically equivalent set of production rules. Let

$$P \equiv \{b_i \mapsto t_i | 0 \le i < n\}$$

be such a set.

Notations and Definitions

For an arbitrary PR p, $p.g$ and $p.a$ denote the guard and the assignment of p, respectively. The predicate $R(a)$, the *result* of the simple assignment a, is defined as: $R(x \uparrow) = x$, and $R(x \downarrow) = \neg x$. An execution of a PR that changes the value of the assigned variable is called *effective*, otherwise it is called *vacuous*.

With these definitions, the stability of a PR can be reformulated as follows:

Stability. *A PR p is stable in a computation if and only if $p.g$ can be falsified only in states where $R(p.a)$ holds. As a consequence, $p.g$ holds as a postcondition of any effective firing of p.*

The production-rule expansion algorithm compiles a handshaking expansion S into a set P of PRs, all of which are stable, with the exception of those whose guards contain negated probes. Since, as we shall see, the guards of the PRs are obtained by strengthening the wait-conditions of S, the stability of the wait-conditions is necessary to satisfy the stability of the PRs.

A wait-condition w is stable if once w is **true** , it remains **true** at least until the completion of the following assignment. Unstable wait-conditions can be caused by negated probes or unrestricted shared variables. The case of negated probes will be dealt with separately by introducing synchronizers. We ignore the use of shared variables in these lecture notes.

In particular, the wait-conditions of the handsaking expansions are stable, also after reshuffling.

Sequencing

The set P of PRs implements S when the following conditions are fulfilled:
Guard strengthening The guards of the PRs of P are obtained by strengthening the wait conditions of S: $\forall i :: b_i \Rightarrow w_i$ and, in the initial state, $w_0 \Rightarrow b_0$.
Sequential execution $(\mathbf{N}i :: b_i \wedge \neg R(t_i)) \leq 1$, i.e., at most one effective PR can be executed at a time.
Program-order execution For all i: If w_{i+1} holds eventually as a postcondition of t_i in S, then b_{i+1} holds eventually as a postcondition of t_i in P. (Addition $i + 1$ is modulo n.)

The first condition establishes that an execution of PR $b_i \mapsto t_i$ in P is equivalent to an execution of $[w_i]; t_i$ in S. The second and third conditions establish that the order of execution of effective PRs of P is the order specified by S, which we have called the *program-order*, and that no deadlock is introduced in the construction of P.

As we shall see, it is not always possible to construct, for a given handshaking expansion, a PR set that satisfies the above three conditions. In certain cases, the handshaking expansion must be augmented with assignments to new variables, called *state variables*. This transformation, which is always possible, will be explained later.

Acknowledgement

Fulfilling the second and third conditions requires that for any two PRs $p :$ $b \mapsto t$ and $p' : b' \mapsto t'$, such that p immediately precedes p' in the program order,

$$b' \Rightarrow R(t)$$

holds as a postcondition of p. We say that b' is the *acknowledgement* of t. Hence the

Acknowledgement Property. *For a PR set executed in program order, the guard of each PR is an acknowledgement of the immediately preceding assignment.*

We shall see that the acknowledgement property is necessary but not sufficient to ensure program-order execution.

We use two kinds of acknowledgements depending on the type of variable used in the assignment. But other forms of acknowledgments can be envisioned. If t assigns an internal variable, then the acknowledgement is implemented by strengthening b' as $b' \wedge R(t)$.

For example, if t is $x \uparrow$, the acknowledgement is $b' \wedge x$.

If t assigns a handshake variable, i.e., a variable implementing a communication command, another kind of acknowledgement can be used as follows. **Acknowledgement of Output Variables.** *For xo and xi used in an active protocol, xi is an acknowledgment of $xo \uparrow$, $\neg xi$ is an acknowledgment of $xo \downarrow$. For xo and xi used in a lazy-active protocol, xi is an acknowledgment of $xo \uparrow$. For yo and yi used in the passive protocol of 6.7, $\neg yi$ is an acknowledgment of $yo \uparrow$, yi is an acknowledgment of $yo \downarrow$.*

Implementation of Stability

Consider a PR set P, which implements a given program S. We are going to show that the acknowledgement property, which is necessary to construct a P that implements S, is also sufficient to guarantee stability.

The execution of a PR p of P establishes a path between a constant node (either VDD or GND), and the node implementing the variable—say, x—assigned by p. Either $p.g$ holds forever after p; or the firing of another PR I, the *invalidating* PR of p, will establish $\neg p.g$, hence cutting the path from the constant node to x.

Let $\tilde{p}$ be the complementary PR of p, i.e., the PR with the complementary assignment. If the PR set contains both p and $\tilde{p}$, then it also contains I because of the non-interference requirement between complementary PRs. And we have the order of execution:

$$p \preceq I \prec \tilde{p} \ .$$

In all the states between I and $\tilde{p}$, the original path to x is cut. In that case, we have to see to it that the assignment to x is completed before the path is cut. Hence the

Completion requirement. *Assignment $p.a$ is completed when a PR q is completed whose guard is an acknowledgement of $p.a$. The execution order of the PR set must satisfy*

$$p \prec q \preceq I \ .$$

Since this requirement is already implied by the acknowledgement property, the construction of P automatically guarantees stability.

We can add an extra requirement to eliminate the pathological cases of "disguised" self-invalidating PRs, even though such cases rarely arise in practice, and they can be dealt with at the implementation level.

Self-Invalidating PRs

Definition. *A PR p is self-invalidating when $R(p.a) \Rightarrow \neg p.g$.*

For example, $\neg x \mapsto x \uparrow$ is self-invalidating.

Self-invalidating PRs are disallowed since they violate the stability requirement. Fortunately, they are excluded by the completion requirement since it implies $I \neq p$.

For instance, the circuit consisting of an inverter with its output connected to its input is excluded by the completion requirement since it corresponds to the PR set:

$$\neg x \mapsto x \uparrow$$
$$x \mapsto x \downarrow$$

and the two PRs of the set are self-invalidating. However, the PR set

$$\neg x \mapsto y \uparrow$$
$$y \mapsto x \uparrow$$
$$x \mapsto y \downarrow$$
$$\neg y \mapsto x \downarrow$$

fulfils the completion requirement, although it is the same circuit as previously, since the only change is the addition of the wire $y \underline{w} x$.

We eliminate such "disguised" self-invalidating PRs by adding the

Restoring Acknowledgement Requirement. *There is at least one restoring PR r satisfying $p \prec r \preceq I$, where r is restoring if it is not part of a wire or a fork.*

With this extra requirement, all forms of self-invalidating PRs are eliminated.

It is remarkable that the acknowledgement requirement, which is necessary to enforce the sequential execution of a PR set, is also sufficient to satisfy stability. From now on, we can manipulate PRs as if the transitions were discrete. However, we have made no simplifying assumption on the physical behavior of the system. The only physical requirement so far is that of monotonicity.

Another requirement on the implementation is that the rings of operators that constitute a circuit keep oscillating. It turns out that eliminating self-invalidating PRs enforces the condition that a ring contain at least three restoring operators, which is a necessary (and in practice also sufficient) condition for the ring to oscillate, thanks to the "gain" property of restoring gates. (See [13] for an explanation of gain.).

6.4.6 Example 1: The (L/R) process

As a first example, let us implement the simple process (L/R), where R is an active channel. This process is one of the basic building blocks for implementing sequencing. The handshaking expansion gives:

$$*[[li]; \ ro\uparrow; \ [ri]; \ ro\downarrow; \ [\neg ri]; \ lo\uparrow; \ [\neg li]; \ lo\downarrow] \ . \tag{6.8}$$

We now consider the handshaking expansion as the specification of the implementation: Any implementation of the program has to satisfy the ordering defined by 6.8. The next step is to construct a production-rule set that satisfies this ordering. We start with the production-rule set that is syntactically derived from 6.8:

$$li \mapsto ro\uparrow$$
$$ri \mapsto ro\downarrow$$
$$\neg ri \mapsto lo\uparrow$$
$$\neg li \mapsto lo\downarrow$$

(As a clue to the reader, PRs of a set are listed in program order.)

Since the program is deadlock-free, effective execution of the PRs in program order is always possible. However, some other execution orders may also be possible. If execution orders other than the program order are possible for the production-rule set, the guards of some rules are strengthened so as to eliminate these execution orders.

In our example, program order is not the only execution order for the syntactic production-rule set: Since $\neg ri$ holds initially, the third PR can be executed first. This is also true for the fourth PR; but the execution of the fourth rule in the initial state is vacuous. Because all handshaking variables of R are back to **false** when R is completed, we cannot find a guard for the transition $lo\uparrow$ that holds only as a precondition of $lo\uparrow$ in 6.8. Hence, we cannot distinguish the state following R from the state preceding R, and thus the sequential execution condition introduced in section 8 cannot be satisfied.

This is a general problem, since it arises for each unshuffled communication action. In order to fulfil the sequential execution condition, we have to guarantee that each state of the handshaking expansion is unique, i.e., there exists a predicate in terms of variables of the program that holds only in this state. The task of transforming the handshaking expansion so as to make each state unique is called *state assignment*.

State Assignment With State Variables

The first technique to define uniquely the state in which the transition $lo\uparrow$ is to take place consists in introducing a state variable, say x, initially **false**. Handshaking expansion 6.8 becomes

$$*[[li];\ ro\uparrow;\ [ri];\ x\uparrow;\ [x];\ ro\downarrow;\ [\neg ri];\ lo\uparrow;\ [\neg li];\ x\downarrow;\ [\neg x];\ lo\downarrow]\ . \tag{6.9}$$

Observe that 6.9 is semantically equivalent to 6.8, since the two sequences of actions that are added to 6.8, namely, $x\uparrow;[x]$ and $x\downarrow;[\neg x]$, are equivalent to a **skip**. (The newly introduced variable x is used nowhere else.)

There are several places where the two assignments to the state variable can be introduced. In general, a good heuristic is to introduce those assignments at such places that the alternation between waits and assignments is maintained. But there are other heuristics that can play a role in the placement of the variables.

Once state variables have been introduced so as to distinguish any two states of the handshaking expansion, it is possible to strengthen the guards of the PRs to enforce program-order execution. The basic algorithm for guard strengthening can be found in [11]. We shall not describe it here. Applied to 6.9, it gives

$$\neg x \wedge li \mapsto ro \uparrow$$
$$ri \mapsto x \uparrow$$
$$x \mapsto ro \downarrow$$
$$x \wedge \neg ri \mapsto lo \uparrow$$
$$\neg li \mapsto x \downarrow$$
$$\neg x \mapsto lo \downarrow .$$

It is easy to check that the acknowledgment property is fulfilled and that the only possible execution order for the above production-rule set is the program order defined by 6.9.

6.4.7   Operator Reduction

The last step of the compilation, called *operator reduction*, groups together the PRs that assign the same variables. Those PRs are then identified with (and implemented as) an operator. The program is thus identified with a set of operators.

Since we have enforced the stability of each rule and non-interference between any two complementary rules, we can implement any set of PRs directly. (For reasons of efficiency, we have to see to it that the guards do not contain too many variables in a conjunct, which would lead to too many transistors in series. Hence, the implementation of the set may also involve decomposing a PR into several PRs by introducing new internal variables.)

The direct operator implementation of the PR set is straightforward:

The PRs that set and reset ro correspond to the asymmetric C-element $(\neg x, li)$ $\underline{aC}$ ro.

The PRs that set and reset lo correspond to the asymmetric C-element $(x, \neg ri)$ $\underline{aC}$ lo.

The PRs that set and reset x correspond to the flip-flop (ri, li) $\underline{ff}$ x.

If the above operators are implemented as dynamic, this implementation of process (L/R) is the simplest possible. If static implementations of the operators are required, another implementation might be considered with fewer state-holding elements since, as we have explained in the first part, static state-holding operators are slightly more difficult to realize than combinational operators.

A last transformation, called *symmetrization*, may be performed on the PR set to minimize the number of state-holding operators. However, since symmetrization also introduces inefficiencies of its own, it should not be applied blindly.

6.4.8   Symmetrization

Symmetrization is performed on the two guards of PRs $b1 \mapsto z \uparrow$ and $b2 \mapsto z \downarrow$, when one of the two guards, say, $b1$, is already in the form $x \wedge \neg b2$. If we replace guard $b2$ with $\neg x \vee b2$, then the two guards are complements of each other; i.e., the operator is combinational. Of course, weakening guard $b2$ is a dangerous transformation since it may introduce a new state where the guard holds. We have to check that this does not occur by checking the following invariant:

Given the new rule $\neg x \vee b2 \mapsto z \downarrow$, $\neg z$ must hold in any state where $\neg x \wedge \neg b2$ holds; i.e., we have to check the invariant truth of

$$x \vee b2 \vee \neg z .$$

Operator Reduction of the (L/R)-element

The symmetrization of the PRs of the (L/R)-element gives:

$$\neg x \wedge li \mapsto ro \uparrow$$
$$ri \mapsto x \uparrow$$
$$\neg li \vee x \mapsto ro \downarrow$$
$$x \wedge \neg ri \mapsto lo \uparrow$$
$$\neg li \mapsto x \downarrow$$
$$ri \vee \neg x \mapsto lo \downarrow$$

The PRs that set and reset ro correspond to the *and*-operator $(\neg x, li) \wedge ro$.

The PRs that set and reset x correspond to the flip-flop $(ri, li) \, \underline{f\!f} \, x$.

The PRs that set and reset lo correspond to the *and*-operator $(x, \neg ri) \wedge lo$.

The flip-flop can be replaced with the C-element $(li, ri) \, \underline{C} \, x$.

The resulting circuit is shown in Figure 6.3. (The dot identifies the input that is activated first.) This implementation of (L/R), either with a flip-flop or with a C-element, is called a *Q-element*. The Q-element implementing (L/R) as above is described by the infix notation $(li, lo) \, \underline{Q} \, (ri, ro)$.

6.4.9   Isochronic Forks

In the previous operator reduction, li is an input to the flip-flop $(li, ri) \, \underline{f\!f} \, x$, and to the *and*-operator $(li, \neg x) \wedge ro$. Formally, in order to compose the PRs

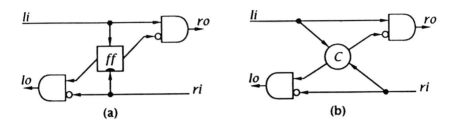

(a)   (b)

Figure 6.3: Implementation of (L/R) with a Q-element

together to form a circuit, we have to introduce the fork $lif(l1, l2)$ and replace li by $l1$ as input of the *and*-operator, and by $l2$ as input of the flip-flop. We also have to introduce the forks $ri \underline{f} (r1, r2)$ and $x \underline{f} (x1, x2)$ for the same reason.

Let us analyse the effect of the first fork only. The PR set that includes the PRs of the fork is:

$$li \mapsto l1 \uparrow, l2 \uparrow$$
$$\neg x \wedge l1 \mapsto ro \uparrow$$
$$ri \mapsto x \uparrow$$
$$\neg l1 \vee x \mapsto ro \downarrow$$
$$x \wedge \neg ri \mapsto lo \uparrow$$
$$\neg li \mapsto l1 \downarrow, l2 \downarrow$$
$$\neg l2 \mapsto x \downarrow$$
$$ri \vee \neg x \mapsto lo \downarrow$$

Now we observe that transition $l1 \uparrow$ o is acknowledged by the guard of the textually following PR but $l2 \uparrow$ is not; transition $l2 \downarrow$ of is acknowledged by the guard of the textually following PR but $l1 \downarrow$ is not. Hence, the assignments $l2 \uparrow$ and $l1 \downarrow$ do not fulfil the completion requirement, and thus are not stable!

We solve this problem by making a simplifying assumption: We assume that the fork is *isochronic*, i.e., *the difference in delays between the two branches of the fork is shorter than the delays in the operators to which the fork is an input.* Hence, when a transition on one output is acknowledged and thus completed, the transition on the other output is also acknowledged and thus completed.

This is the only timing condition that has to be fulfilled. In general, the constraint is easy to meet because it is one-sided. However, the isochronicity requirement is more difficult to meet when a negated input introduces an inverter on a branch of the fork, since the transition delays of an inverter are of the same order of magnitude as the transition delays of other operators. We have proved that, for the implementation of each language construct, these inverters can always be eliminated from the isochronic forks by simple transformations. (These transformations have not been applied to the circuits presented here as examples, but they are always applied before the circuits are actually implemented.)

In [12], we have proved that the class of entirely delay-insensitive circuits is very limited: Practically all circuits of interest fall outside the class. We believe that the notion of isochronic fork is the weakest compromise to delay-insensitivity sufficient to implement any circuit of interest.

Which forks have to be isochronic is easy to decide by a simple analysis of the PR sets. For instance, the fork $ri\underline{f}\,(r1, r2)$ also has to be isochronic, but the fork $x\underline{f}\,(x1, x2)$ does not. We shall ignore the issue of isochronic forks in the rest of this presentation.

6.4.10   Example 2: A one-place buffer

The one-place buffer is the most ubiquitous process. In the processor for example, each stage of the pipeline is a one-place buffer of the type:

$$*[L?x;\, R!f(x)]\ .$$

Let us ignore the transmission of messages, and implement the "bare" process:

$$*[L;\, R]\ .$$

One of the most useful implementations of this process is with L lazy-active and R passive. The handshaking expansion gives:

$$*[[\neg li];\ lo\!\uparrow;\ [li];\ lo\!\downarrow;\ [ri];\ ro\!\uparrow;\ [\neg ri];\ ro\!\downarrow]\ .$$

We choose to include the state variable x in such a way that the transition $x\uparrow$ is concurrent with $lo\!\uparrow$, and transition $x\downarrow$ is concurrent with $ro\!\uparrow$. We get:

$$*[[\neg li];\ lo\!\uparrow;\ x\!\uparrow;\ [x];\ [li];\ lo\!\downarrow;\ [ri];\ ro\!\uparrow;\ x\!\downarrow;\ [\neg x];\ [\neg ri];\ ro\!\downarrow]\ .$$

The production rule expansion is:

$$\neg x \wedge \neg li \wedge \neg ro \mapsto lo \uparrow$$
$$lo \mapsto x \uparrow$$
$$x \wedge li \mapsto lo \downarrow$$
$$x \wedge \neg lo \wedge ri \mapsto ro \uparrow$$
$$ro \mapsto x \downarrow$$
$$\neg x \wedge \neg ri \mapsto ro \downarrow$$

The direct implementation of this production rule set is shown in Figure 6.4.

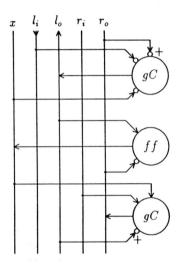

Figure 6.4: A circuit for the one-place buffer

6.4.11   Example 3: Single-Variable Register

Consider the following *register* process that provides read and write access to a simple boolean variable, x:

$$*[[\overline{P} \to P?x$$
$$[\overline{Q} \to Q!x \qquad\qquad (6.10)$$
$$]],$$

where $\neg\overline{P} \vee \neg\overline{Q}$ holds at any time.

The handshaking expansion uses the *double-rail* technique: The boolean value of x is encoded on two wires, one for the value **true** and one for the value **false** . Input channel P has two input wires, $pi1$ for receiving the value **true** , and $pi2$ for receiving the value **false** ; and one output wire, po. Output channel Q has two output wires, $qo1$ for sending the value **true** , and $qo2$ for sending the value **false** ; and one input wire, qi. Each guarded command is expanded to two guarded commands:

$$*[[\, pi1 \rightarrow x \uparrow;\ [x];\ po \uparrow;\ [\neg pi1];\ po \downarrow$$
$$[\![\ pi2 \rightarrow x \downarrow;\ [\neg x];\ po \uparrow;\ [\neg pi2];\ po \downarrow$$
$$[\![\ x \wedge qi \rightarrow qo1 \uparrow;\ [\neg qi];\ qo1 \downarrow \qquad (6.11)$$
$$[\![\ \neg x \wedge qi \rightarrow qo2 \uparrow;\ [\neg qi];\ qo2 \downarrow$$
$$]]\ .$$

Mutual Exclusion Between Guarded Commands

We are now faced with a new problem: enforcing mutual exclusion between the production-rule sets of different guarded commands. (This problem is not concerned with making the *guards* of the different commands mutually exclusive. For the time being, we are considering only examples where the guards of the commands are already mutually exclusive.) Let us illustrate our problem with the compilation of the first two guarded commands. If we just concatenate the production-rule sets of these two commands, we get:

$$pi1 \mapsto x \uparrow$$
$$pi1 \wedge x \mapsto po \uparrow$$
$$\neg pi1 \mapsto po \downarrow$$
$$pi2 \mapsto x \downarrow$$
$$pi2 \wedge \neg x \mapsto po \uparrow$$
$$\neg pi2 \mapsto po \downarrow\ .$$

But we now observe that the second and the sixth guarded commands are interfering (they set and reset the same variable po); and, for reasons of symmetry, the same holds for the third and the fifth PRs.

Hence, the problem of ensuring mutual exclusion between PRs of different guarded commands is the same as enforcing program order between PRs of the same guarded command. We use the same technique, which consists in strengthening the guards of the production rules, if necessary, by introducing state variables to distinguish between the states corresponding to each true guard.

In the case at hand, we strengthen the guards of the third and the sixth rules as:

$$x \wedge \neg pi1 \mapsto po \downarrow$$
$$\neg x \wedge \neg pi2 \mapsto po \downarrow\ .$$

The rest of the implementation is straightforward. The first and fourth PRs correspond to the flip-flop $(pi1, \neg pi2) \underline{ff} \ x$. The other PRs can be transformed into

$$(pi1 \wedge x) \vee (pi2 \wedge \neg x) \mapsto po \uparrow$$
$$(\neg pi1 \wedge \neg x) \vee (\neg pi2 \wedge x) \mapsto po \downarrow ,$$

which is the definition of the *if*-operator $(pi1, pi2, x) \ \underline{if} \ po$.

The production-rule expansion of the last two guarded commands gives:

$$x \wedge qi \mapsto qo1 \uparrow$$
$$\neg x \vee \neg qi \mapsto qo1 \downarrow$$
$$\neg x \wedge qi \mapsto qo2 \uparrow$$
$$x \vee \neg qi \mapsto qo2 \downarrow ,$$

which corresponds to the two operators $(x, qi) \triangle qo1$ and $(\neg x, qi) \triangle qo2$. The circuit is represented in Figure 6.5.

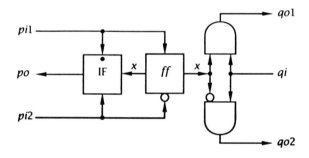

Figure 6.5: Single boolean register

In the next example, we shall refer to the implementation of the first two guarded commands as the *register* operator:

$$(pi1, pi2) \ \underline{reg} \ (po, x) .$$

We shall refer to the implementation of the last two guarded commands of (26) as the *read* operator:

$$(qi, x) \ \underline{read} \ (qo1, qo2) .$$

6.5   Case Study: Distributed Mutual Exclusion

The first paper describing this method for the synthesis of asynchronous circuits from high-level description was presented at the 1985 Chapel Hill Conference on VLSI [8]. The example used to illustrate the method was the algorithm for distributed mutual exclusion on a ring of processes described in Part 1 of these notes.

Unfortunately, the circuit presented in the Chapel Hill paper is not entirely correct: A glitch may appear on the wire named z in the paper. The error is due to my not following the compilation procedure when I defined the variable z. The error was noticed by many people, and the actual CMOS implementation of the circuit realized by Andy Fife the same year is entirely correct.

However, I never took the time to publish the correct solution, and therefore the bug has been rediscovered over and over again, sometimes with great publicity[2]. Since several people have asked me to show them a correct derivation of the circuit, here it is after five years!

As in the original paper, we observe that the two consecutive D commands, and the two consecutive U commands can both be implemented as the two halves of a 4-phase handshaking protocol; and therefore we can replace the two U commands with one single U to be implemented as a 4-phase handshaking protocol.

Next, we decompose process m into two processes A and B as follows:

$$A \equiv *[\,[\,\overline{U} \to P; Bt; U$$
$$|\overline{L} \to P; Bf; L$$
$$]\,]$$

$$B \equiv *[\,[\,\overline{Q} \wedge b \to Q$$
$$|\overline{Q} \wedge \neg b \to R; Q$$
$$|\overline{S} \to b\uparrow; S$$
$$|\overline{T} \to b\downarrow; T$$
$$]\,]$$

The internal channels between A and B are (P, Q), (Bt, S), and (Bf, T).

The technique used to obtain A and B is the standard process decomposition, with one addition. The plain process decomposition would give a process A with U before Bt, and L before Bf. We have inverted the order of these actions, since it is semantically irrelevant whether the assignment to b is the last action of the guarded command provided the assignment follows the selection command. The reason for this transformation is that the program in which U and L are the last actions of the guarded commands is easier to implement. This point will be further explained in the compilation of A.

Compilation of A

Since the guards $\overline{U}$ and $\overline{L}$ are not mutually exclusive, we are introducing an arbiter described by the program:

$$Arb \equiv *[\,[ui \rightarrow u'\uparrow;\ [\neg ui];\ u'\downarrow$$
$$|li\ \rightarrow l'\uparrow;\ [\neg li];\ l'\downarrow$$
$$]\,]\ .$$

We know that $A = (Arb\|A')$, where $A' = A_{u',l'}^{ui,li}$.
EXERCISE Prove the correctness of the above result. □
   Hence:

$$A' \equiv *[\,[u' \rightarrow po\uparrow;\ [pi];\ po\downarrow;\ [\neg pi];\ bto\uparrow;\ [bti];\ bto\downarrow;\ [\neg bti];\ uo\uparrow;\ [\neg u'];\ uo\downarrow$$
$$\llbracket l' \rightarrow po\uparrow;\ [pi];\ po\downarrow;\ [\neg pi];\ bfo\uparrow;\ [bfi];\ bfo\downarrow;\ [\neg bfi];\ lo\uparrow;\ [\neg l'];\ lo\downarrow$$
$$]\,]$$

Mutual exclusion among guarded commands

The main problem in implementing A' is to enforce the mutual exclusion between the two guarded commands (GCs). By construction of the arbiter circuit Arb, we know that—provided that $\neg u' \wedge \neg l'$ holds initially—$\neg u' \vee \neg l'$ holds at any time. Hence, the mutual exclusion between the *guards* of A' is guaranteed.

However, as soon as $u'\downarrow$ is completed, the first GC of the arbiter can complete, the second GC of the arbiter can start, and consequently, the PR set implementing the second GC of A' can start firing, even though the first GC of A' may not be completed. We shall see that, in order to enforce the mutual exclusion between the implementations of the two GCs of A', it is advantageous to postpone $u'\downarrow$ as long as possible. This explains our decision to modify A such that U is the last action of the first GC, and L the last action of the second GC.

6.5.1   First Solution

We slightly reshuffle the actions of A' as follows:

$$A' \equiv *[\,[u' \rightarrow po\uparrow;\ [pi];\ po\downarrow;\ [\neg pi];\ bto\uparrow;\ [bti];\ uo\uparrow;\ [\neg u'];\ bto\downarrow;\ [\neg bti];\ uo\downarrow$$
$$\llbracket l' \rightarrow po\uparrow;\ [pi];\ po\downarrow;\ [\neg pi];\ bfo\uparrow;\ [bfi];\ lo\uparrow;\ [\neg l'];\ bfo\downarrow;\ [\neg bfi];\ lo\downarrow$$
$$]\,]$$

We first ignore the transitions on *bto*, *bti*, *bfo*, and *bfi*, and implement the program:

$$A' \equiv *[\,[u' \rightarrow po\uparrow;\ [pi];\ po\downarrow;\ [\neg pi];\ uo\uparrow;\ [\neg u'];\ uo\downarrow$$
$$[l' \rightarrow po\uparrow;\ [pi];\ po\downarrow;\ [\neg pi];\ lo\uparrow;\ [\neg l'];\ lo\downarrow$$
$$]\,]$$

Each guarded command is a Q-element. The transitions on *bto*, *bti*, *bfo*, and *bfi* are added by just "opening" the wires *uo* and *lo*, respectively.

For mutual exclusion between the implementations of the two guarded commands, the guard u' is strengthened as $u' \wedge \neg lo$, and the guard l' is strenghened as $l' \wedge \neg uo$.

Merge

We now have to compose the circuit implementing the first GC with the one implementing the second GC. This composition is a little more than mere juxtaposition because the two circuits use the variables *pi* and *po*. The standard way to deal with this case is to compose the two circuits with a *merge* circuit.

We replace P with $P1$ in the first GC, and with $P2$ in the second GC, and add the *merge* process:

$$*[\,[\overline{P1} \rightarrow P1 \bullet P$$
$$[\overline{P2} \rightarrow P2 \bullet P$$
$$]\,]$$

The handshaking expansion gives:

$$*[\,[p1i \rightarrow po\uparrow;\ [pi];\ p1o\uparrow;\ [\neg p1i];\ po\downarrow;\ [\neg p1i];\ p1o\downarrow$$
$$[p2i \rightarrow po\uparrow;\ [pi];\ p2o\uparrow;\ [\neg p2i];\ po\downarrow;\ [\neg p2i];\ p2o\downarrow$$
$$]\,]$$

The production rule expansion gives:

$$p1i \vee p2i \mapsto po\uparrow$$
$$pi \wedge p1i \mapsto p1o\uparrow$$
$$\neg p1i \wedge \neg p2i \mapsto po\downarrow$$
$$\neg pi \mapsto p1o\downarrow$$
$$pi \wedge p2i \mapsto p2o\uparrow$$
$$\neg pi \mapsto p2o\downarrow\ .$$

The operators are the or-gate $(p1i, p2i)\ \underline{\vee}\ po$, and the two asymmetric C-elements $(pi; p1i)\ \underline{aC}\ p1o$ and $(pi; p2i)\ \underline{aC}\ p2o$.

Circuit for A'

Composing the merge circuit and the circuits for the two guarded commands lead to an implementation of A'. But we make two observations. First, the asymmetric C-elements in the merge are not needed in this case. Second, and more importantly, we realize that instead of merging the two circuits after the two Q-elements, we could merge them before the Q-elements so that the two circuits could share the same Q-element. This transformation is formalized by the following program decomposition. We have $A' \equiv (A1\|Q)$, with:

$$A1 \equiv *[\quad [u' \wedge \neg lo \quad \rightarrow po' \uparrow; \ [pi']; \ bto \uparrow; [bti]; \ uo \uparrow; [\neg u'];$$
$$po' \downarrow; \ [\neg pi']; \ bto \downarrow; [\neg bti]; \ uo \downarrow$$
$$[\![l' \wedge \neg uo \quad \rightarrow po'' \uparrow; \ [pi']; \ bfo \uparrow; \ [bfi]; \ lo \uparrow; [\neg l'];$$
$$po'' \downarrow; \ [\neg pi']; \ bfo \downarrow; [\neg bfi]; \ lo \downarrow$$
$$] \ ]$$

$$Q \equiv *[[po' \vee po'']; \ po \uparrow; \ [pi]; \ po \downarrow; \ [\neg pi]; \ pi' \uparrow; \ [\neg po' \wedge \neg po'']; \ pi' \downarrow]$$

The first guarded command of $A1$ is compiled as:

$$u' \wedge \neg lo \mapsto po' \uparrow$$
$$pi' \wedge po' \mapsto bto \uparrow$$
$$bti \mapsto uo \uparrow$$
$$lo \vee \neg u' \mapsto po' \downarrow$$
$$pi' \mapsto bto \downarrow$$
$$\neg bti \mapsto uo \downarrow$$

The operator reduction gives:

$$(u', \neg lo) \underline{\wedge} po'$$
$$(pi'; po') \underline{aC} bto$$
$$bti \ \underline{w} \ uo$$

The compilation of the second GC of $A1$ is similar.

Compilation of B

The compilation of B is identical to that of the original paper. The hand-shaking expansion of B with a slight reshuffling of the actions in the second GC gives:

$$B \equiv *[[qi \wedge b \rightarrow qo \uparrow; [\neg qi]; qo \downarrow$$
$$[\![qi \wedge \neg b \rightarrow ro \uparrow; [ri]; qo \uparrow; [\neg qi]; ro \downarrow; [\neg ri]; qo \downarrow$$
$$[\![si \rightarrow b \uparrow; so \uparrow; [\neg si]; so \downarrow$$
$$[\![ti \rightarrow b \downarrow; to \uparrow; [\neg ti]; to \downarrow$$
$$]]$$

We first observe that the mutual exclusion between the guards and between the guarded commands is guaranteed. The production rule expansion gives:

$$qi \wedge b \mapsto qo \uparrow$$
$$b \wedge \neg qi \mapsto qo \downarrow$$
$$qi \wedge \neg b \mapsto ro \uparrow$$
$$ri \mapsto qo \uparrow$$
$$\neg qi \mapsto ro \downarrow$$
$$\neg b \wedge \neg ri \mapsto qo \downarrow$$

The conjunct b is added to the guard of the first PR for mutual exclusion with the second GC. A better strengthening of the two rules that reset qo is $\neg qi \wedge \neg ri \mapsto qo \downarrow$.

Combining all PRs relative to qo gives:

$$qi \wedge b \vee ri \mapsto qo \uparrow$$
$$\neg qi \wedge \neg ri \mapsto qo \downarrow$$

The other operator is $(qi; \neg b)\ \underline{aC}\ ro$. The production rule expansion of the last two GCs is straightforward. It gives:

$$si \mapsto b \uparrow$$
$$ti \mapsto b \downarrow$$
$$si \wedge b \mapsto so \uparrow$$
$$\neg si \mapsto so \downarrow$$
$$ti \wedge \neg b \mapsto to \uparrow$$
$$\neg ti \mapsto to \downarrow$$

The set of operators is:

$$(si; \neg ti)\ \underline{f\!f}\ b$$
$$(si; b)\ \underline{aC}\ so$$
$$(ti; \neg b)\ \underline{aC}\ to$$

The last two operators can be replaced with the and-gates $(si, bi) \wedge so$ and $(ti, \neg b) \wedge to$ by the usual symmetrization. The flip-flop can be replaced with the C-element $(si, \neg ti)\ \underline{C}\ b$, also by symmetrization.

The complete circuit is shown in Figure 6.6.

6.5.2 Exercise: Implementation without reshuffling

Can we implement the program of A directly without postponing U and L? We have to implement the following version of A:

$$A \equiv *[\,[\overline{U} \rightarrow P; U; Bt$$
$$[\overline{L} \rightarrow P; L; Bf$$
$$]\,]$$

B is unchanged.

$$A1 \equiv (Arb\|A'),$$

where Arb is unchanged. A' is slightly reshuffled.

$$A' \equiv *[\,[u' \rightarrow po \uparrow; [pi]; uo \uparrow; [\neg u']; po \downarrow; [\neg pi]; uo \downarrow; bto \uparrow; [bti]; bto \downarrow; [\neg bti]$$
$$\|l' \rightarrow po \uparrow; [pi]; lo \uparrow; [\neg l']; po \downarrow; [\neg pi]; lo \downarrow; bfo \uparrow; [bfi]; bfo \downarrow; [\neg bfi]$$
$$]\,]$$

Apart from the opening of the uo wire for the (po, pi) connection, the first guarded command is just the passive-active buffer:

$$*[[u']; uo \uparrow; [\neg u']; uo \downarrow; bto \uparrow; [bti]; bto \downarrow; [\neg bti]]$$

The rest of the compilation is left as an exercise to the reader.

Acknowledgments

I am indebted to my students Dražen Borković, Steve Burns, Marcel van der Goot, Pieter Hazewindus, Tony Lee, and José Tierno for their many contributions to the research.

6.6   References

[1] Steven M. Burns and Alain J. Martin. Syntax-directed Translation of Concurrent Programs into Self-timed Circuits. *Proc. Fifth MIT Conference on Advanced Research in VLSI*, ed. J. Allen and F. Leighton, MIT Press, 35-40, 1988.

[2] David L. Dill. *Theory for Automatic Hierarchical Verification of Speed-Independent Circuits.* MIT Press, 1989.

[3] K. Mani Chandy and Jayadev Misra. *Parallel Program Design: A Foundation.* Addison-Wesley, Reading MA, 1988.

[4] Edsger W. Dijkstra. *A Discipline of Programming.* Prentice-Hall, Englewood Cliffs NJ, 1976.

[5] C.A.R. Hoare. Communicating Sequential Processes. *Comm. ACM* 21,8, pp 666-677, 1978.

[6] Alain J. Martin. The Probe: An Addition to Communication Primitives. *Information Processing letters* 20, pp 125-130, 1985.

[7] Alain J. Martin. Distributed Mutual Exclusion on a Ring of Processes. *Science of Computer Programming*, 5, 265-276, 1985.

[8] Alain J. Martin. The Design of a Self-timed Circuit for Distributed Mutual Exclusion. *1985 Chapel Hill Conference on VLSI*, ed. Henry Fuchs, Computer Science Press, 247-260, 1985.

[9] A.J. Martin, S.M. Burns, T.K. Lee, D. Borkovic, P.J. Hazewindus. The Design of an Asynchronous Microprocessor. *Decennial Caltech Conference on VLSI*, ed. C.L. Seitz, MIT Press, 351-273, 1989.

[10] Alain J. Martin. Compiling Communicating Processes into Delay-insensitive VLSI circuits. *Distributed Computing*, 1,(4), 1986.

[11] Alain J. Martin. Formal Program Transformations for VLSI Circuit Synthesis. *UT Year of Programming Institute on Formal Developments of Programs and Proofs*, ed. E.W. Dijkstra, Addison-Wesley, Reading MA, 1989.

[12] Alain J. Martin. The Limitations to Delay-Insensitivity in Asynchronous Circuits. *Sixth MIT Conference on Advanced Research in VLSI*, ed. W.J. Dally, MIT Press, 1990.

[13] Carver Mead and Lynn Conway. *Introduction to VLSI Systems*, Addison-Wesley, Reading MA, 1980.

[14] Teresa H. Meng, Robert W. Brodersen, David G. Messerchmitt. Automatic Synthesis of Asynchronous Circuits from High-Level Specifications. *IEEE Trans. on CAD*, 8:11, 1185-1205, 1989.

[15] Raymond E. Miller. *Switching Theory*, Vol. 2, Wiley, 1965.

[16] J. Staunstrup and M.R. Greenstreet. Designing Delay-Insensitive Circuits using "Synchronized Transitions." *IMEC IFIP International Workshop on Applied Formal Methods for Correct VLSI Design*, 1989.

[17] Charles L. Seitz. System Timing. Chapter 7 in Mead & Conway, *Introduction to VLSI Systems*, Addison-Wesley, Reading MA, 1980.

[18] N. Weste and K. Eshraghian. *Principles of CMOS VLSI Design*, Addison-Wesley Reading MA, 1985.

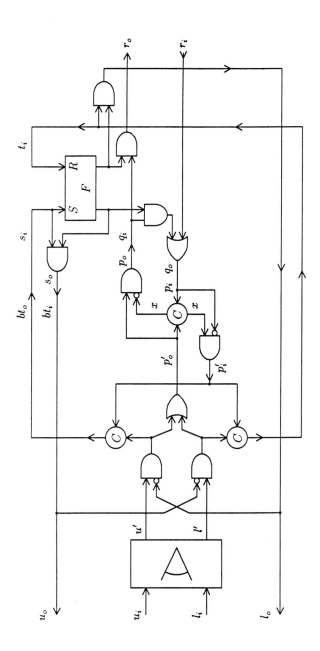

Figure 6.6: Circuit for a server

Formal Methods for VLSI Design
J. Staunstrup (Editor)
Elsevier Science Publishers B.V. (North-Holland)
© IFIP, 1990

Chapter 7

A Formal Introduction to a Simple HDL

Bishop C. Brock and Warren A. Hunt, Jr. [1]

Abstract. A hierarchical, occurrence-oriented, combinational hardware description language has been formalized. Instead of using logic formulas to represent circuits, we represent circuits as list constants. Using a formal logic, interpreters have been defined which give meanings to circuit constants, and a good-circuit predicate recognizes well-formed circuit descriptions. We can directly verify circuit specifications, but instead we often verify functions which generate circuit constants.

7.1 Introduction

The formalization of a hierarchical, occurrence-oriented, combinational hardware description language (HDL) has been accomplished using the Boyer-Moore logic. Circuits are represented as Boyer-Moore list constants, and the Boyer-Moore logic [2] is used to define the semantics and syntax of our circuit constants. Instead of verifying each circuit directly, we often prove the correctness of functions which synthesize circuit constants. We employ the Boyer-Moore theorem prover to mechanically manage our database of definitions and to check our proofs.

CAD vendors have long provided tools which operate on circuit descriptions. The typical paradigm is to record design information in a computer data

---

[1] Computational Logic, Inc., 1717 West Sixth Street, Suite 290, Austin, Texas 78703-4776, USA. e-mail: brock@CLI.COM, hunt@CLI.COM

file and write programs to manipulate this data. Programs often translate data from one form to another, e.g., from a gate-graph to a transistor wiring list. For logistical reasons, the formal methods community has not previously attempted to represent circuits as data nor to verify circuits represented as data; circuits have been modeled as terms in a formal logic [10, 19, 3, 7, 12, 18]. Representing circuits as data involves the definition of interpreters which provide the semantics for the circuit data. To prove the correctness of circuits represented as data requires proofs about the interpretations of the data.

Previously, we used Boyer-Moore logic expressions to represent hardware circuits. Circuits were verified by proving that they satisfied some more abstract specification. However, this approach does not provide a direct migration path to CAD languages, from which an actual physical device can be realized, unless the modeling language is a CAD language itself. Although it may seem that just using an existing CAD language would provide both the modeling capability and an implementation path, commercial CAD languages do not have formal semantics, which means that circuit verification is impossible. Our approach here is to formalize a subset of a conventional CAD language. This approach provides a formal circuit semantics, a formal circuit syntax, and a means of translating circuit descriptions from/to a CAD language.

Here we describe a formalization of a combinational hardware description language and show how we verify functions which generate correct circuit descriptions. Our HDL definition formalizes the notion of circuit delay, fanout, logical values, circuit loading, circuit modules, and circuit module hierarchy; these issues cannot be formally addressed if circuits are modeled with logic formulas. The power of being able to reason about circuits expressed as data cannot be over-emphasized. For instance, we have proved the correctness of a function which produces ALU circuits. We know in advance that any circuit constructed by this ALU-producing function is correct. In addition, modeling circuit specifications with data even admits the possibility of verifying tools (e.g., minimizers, tautology checkers, etc.) which manipulate circuit expressions.

Our presentation begins with an introduction to the notion of hardware verification. We then describe the concept of functions which generate provably correct circuits, at which point the reader should understand the concepts that the rest of the paper makes more precise.

7.2   Hardware Verification

We consider hardware verification to mean the use of formal methods for specifying and verifying the operation of digital computing devices. The hardware verification community is attempting to formalize as much of the digital design process as possible. Presently, engineers are given or create specifications that contain a mixture of formal and informal notations from which they are expected to create working devices. The "hardware verification" approach advocates the use of formal logic for both designs and high-level specifications. We introduce this type of approach here.

The use of formal techniques to design hardware is spreading (for example [12, 3, 4, 5, 6, 18, 11, 14, 17]). Our effort has been ongoing since 1985, and our long-term goal is to provide a means whereby circuits may be rigorously specified and mechanically verified. We conceptualize this notion as providing a mathematical statement, which we call a *formula manual*, that completely specifies the operation of a hardware component. To visualize this notion, imagine a microprocessor user's manual containing a series of formulas that describes the programming model, the timing diagrams, the memory interface, the pin-out, the power requirements, cooling requirements, etc. Further, imagine that available implementations of this microprocessor were verified to meet every specification contained in the formula manual. Then the formula manual would represent a formal specification that would allow hardware engineers to connect this device with complete confidence and would allow software engineers to completely predict the results of programming it.

Formula manuals are a long way off. To address the difficulties of designing, specifying, and constructing real computing equipment, the hardware verification community must continually expand its modeling efforts to explicitly include all hardware attributes which contribute to the correct operation of hardware devices. This expansion will eventually enable us to provide formula manuals for many hardware devices. Formal hardware specification and verification efforts have primarily concentrated on the logical correctness of circuit designs. With the formalization of an HDL, we are expanding our formal model to explicitly model (with varying degrees of precision) circuit fanout, loading, and circuit hierarchy. Our desire to explicitly model more hardware characteristics is part of what drove us to consider the specification of an HDL.

Verified hardware design specifications must be convertible to a form that results in physical devices. Formal hardware design specifications are usually described as netlists of transistors or Boolean gates. This level of description is far removed from actual physical implementation; therefore, the conversion of the formal hardware design specifications into physical descriptions is an area where errors can easily be made. We have designed the HDL presented here to be structured just like some of the HDLs used in commercial CAD systems. It is important that the conversion from verified designs to CAD systems be made at the lowest level possible and be made in a simple and believable way. This is another factor that influenced the design of our HDL.

7.3   The Boyer-Moore Logic

The HDL we have defined is expressed in terms of Boyer-Moore list constants. We use the Boyer-Moore logic to recognize well-formed HDL expressions and to provide a semantics for our HDL. Here we give a quick introduction to the Boyer-Moore logic and present some examples of its use.

The Boyer-Moore logic [2] is a quantifier-free, first-order predicate calculus with equality. Recursive functions may be defined, provided they terminate. Logic formulas are written in a prefix-style, Lisp-like notation. The basic logic includes several built-in data types: Booleans, natural numbers, lists, literal atoms, and integers. Additional data types can be defined.

An unusual feature of the Boyer-Moore logic is the ability to extend the logic by the application of any of the following axiomatic acts: defining conservative functions, adding recursively constructed data types, and adding arbitrary axioms. Adding an arbitrary formula as an axiom does not guarantee the soundness of the logic; we do not use this feature.

The Boyer-Moore theorem prover is a Common Lisp [13] program that provides a user with various commands to extend the logic and to prove theorems. A theorem prover user enters commands through the top-level Common Lisp interpreter. The theorem prover manages the axiom database, user definitions and data types, and proved theorems, thus allowing a user to concentrate on the less mundane aspects of proof development. The theorem prover contains decision procedures for propositional logic and linear arithmetic, and it includes a simplifier and rewriter. The theorem prover also contains procedures for automatically performing structural inductions.

We use the Boyer-Moore theorem prover as a proof checker. We lead the theorem prover to difficult theorems by providing it with a graduated sequence of more and more difficult lemmas until a final result can be obtained.

7.3.1   Bit-Vectors

We represent a bit-vector as a list of Boolean elements. We formalize this notion with the Boyer-Moore logic by defining the functions BOOLP and BVP. In this presentation we write definitions with the "=" symbol, while theorems are presented without the "=" symbol. BOOLP tests that X is either T (true) or F (false).[2] BVP has been defined to recognize a (possibly empty) list of Boolean values. Lists are formed with the pairing function CONS; CAR selects the first element of a pair and CDR the second. For instance, (LIST T F F) is a three-element bit-vector. BVP works recursively; either X is equal to NIL, or (CAR X) is Boolean and (CDR X) is a bit vector recognized by BVP.

---

[2] The semantics for OR, EQUAL, T, and F are described in Boyer and Moore's book, *A Computational Logic Handbook* [2]; see this book for a complete introduction to the Boyer-Moore logic and theorem prover.

```
(BOOLP X) = (OR (EQUAL X T) (EQUAL X F))

(BVP X) = (IF (NOT (LISTP X))
              (EQUAL X NIL)
              (AND (BOOLP (CAR X))
                   (BVP (CDR X))))

(BITN N LIST) = (IF (ZEROP N)
                    (IF (LISTP LIST)
                        (IF (CAR LIST) T F)
                        F)
                    (BITN (SUB1 N) (CDR LIST)))
```

To access a bit in a bit vector we use the function BITN. The Nth bit of LIST is returned if this bit is Boolean; otherwise, F is returned. We will use this formalization of Booleans and bit vectors throughout this paper.

The function APPEND is used to append two list. We can prove that appending two bit vectors together produces a bit vector, and a lemma stating this fact is shown below the definition for APPEND.

```
(APPEND X Y) = (IF (NOT (LISTP X))
                   Y
                   (CONS (APPEND (CAR X) Y)))

(IMPLIES (AND (BVP X)
              (BVP Y))
         (BVP (APPEND X Y)))
```

The proof is by induction on X, and the Boyer-Moore theorem prover can automatically perform this proof.[3]

7.3.2   Boyer-Moore List Constants and Basic Definitions

Our hardware circuit boxes are represented with Boyer-Moore list constants. Constants are written using the Lisp quote notation. The following statements are theorems.

```
(EQUAL (CAR (CONS X Y) X))
(EQUAL (CDR (CONS X Y) Y))

(EQUAL (LISTP (CONS X Y)) T)

(EQUAL '(A B C ... X) (CONS 'A '(B C ... X)))
```

---

[3] The definitions of APPEND and many other simple functions are standard in the Boyer-Moore theorem prover.

```
(EQUAL (CAR '(A B C) 'A))
(EQUAL (CDR '(A B C) '(B C)))

(EQUAL (LIST A B C ... X) (CONS A (LIST B C ... X)))
```

'(A B C) is a list of three literal atoms. We use nested lists to provide a structure for our circuit descriptions; components are accessed using combinations of CARs and CDRs. CAR/CDR nests are abbreviated; for example, we write (CAR (CDR (CDR X))) as (CADDR X).

Included below are functions used repeatedly throughout the remainder of this paper. These definitions should be skipped upon a first reading, and referred to as needed. As with the definitions above, we consider these definitions to be "obviously correct," that is, we use these functions without proof. NLISTP is a predicate which returns true if X is not a list. BOOLFIX coerces X to a Boolean value. FIRSTN collects the first N bits of a list. RESTN collects bits starting at position N in L. ASSOC searches for the key X in association list ALIST and returns a key-value pair or F. COLLECT-ASSOC returns a list of values for the keys in ARGS. MEMBER tests whether X is an element of LIST. DISJOINT is true if no member of L1 is a member of L2. DUPLICATES? returns true if no member of L occurs twice in L. The length of a list is given by LENGTH.

```
(NLISTP X) = (NOT (LISTP X))

(BOOLFIX X) = (IF X T F)

(FIRSTN N L) = (IF (LISTP L)
                   (IF (ZEROP N)
                       NIL
                       (CONS (CAR L) (FIRSTN (SUB1 N) (CDR L))))
                   NIL)

(RESTN N L) = (IF (LISTP L)
                  (IF (ZEROP N)
                      L
                      (RESTN (SUB1 N) (CDR L)))
                  L)
```

```
(ASSOC X ALIST) = (IF (NLISTP ALIST)
                      F
                      (IF (EQUAL X (CAAR ALIST))
                          (CAR ALIST)
                          (ASSOC X (CDR ALIST)))))

(COLLECT-ASSOC ARGS ALIST) = (IF (NLISTP ARGS)
                                 NIL
                                 (CONS (CDR (ASSOC (CAR ARGS) ALIST))
                                       (COLLECT-ASSOC (CDR ARGS)
                                                      ALIST)))

(MEMBER X LIST) = (IF (NLISTP LIST)
                      F
                      (IF (EQUAL X (CAR LIST))
                          T
                          (MEMBER X (CDR LIST)))))

(DISJOINT L1 L2) = (IF (NLISTP L1)
                       T
                       (AND (NOT (MEMBER (CAR L1) L2))
                            (DISJOINT (CDR L1) L2)))

(DUPLICATES? L) = (IF (NLISTP L)
                      F
                      (OR (MEMBER (CAR L) (CDR L))
                          (DUPLICATES? (CDR L))))

(LENGTH X) = (IF (NLISTP X) 0 (ADD1 (LENGTH (CDR X))))

(PAIRLIST L1 L2) = (IF (NLISTP L1)
                       NIL
                       (CONS (CONS (CAR L1) (CAR L2))
                             (PAIRLIST (CDR L1) (CDR L2))))

(PROPERP X) = (IF (NLISTP X) (EQUAL X NIL) (PROPERP (CDR X)))

(SUBSET L1 L2) = (IF (NLISTP L1)
                     T
                     (AND (MEMBER (CAR L1) L2)
                          (SUBSET (CDR L1) L2)))
```

```
(UNION L1 L2) = (IF (LISTP L1)
                    (IF (MEMBER (CAR L1) L2)
                        (UNION (CDR L1) L2)
                        (CONS (CAR L1)
                              (UNION (CDR L1) L2))))
                L2)

(MAX-MEMBER LIST) = (IF (NLISTP LIST)
                        0
                        (MAX (CAR LIST)
                             (MAX-MEMBER (CDR LIST))))

(MAKE-BALANCED-TREE N)
 =
(IF (ZEROP N)
    0
    (IF (EQUAL N 1)
        0
        (CONS (MAKE-BALANCED-TREE (QUOTIENT N 2))
              (MAKE-BALANCED-TREE (DIFFERENCE N (QUOTIENT N 2)))))))

(TREE-SIZE TREE) = (IF (NLISTP TREE)
                       1
                       (PLUS (TREE-SIZE (CAR TREE))
                             (TREE-SIZE (CDR TREE))))

(TFIRSTN LIST TREE) = (FIRSTN (TREE-SIZE (CAR TREE)) LIST)

(TRESTN LIST TREE) = (RESTN (TREE-SIZE (CAR TREE)) LIST)
```

PAIRLIST creates an association list. PROPERP checks that a list ends with NIL. If the members of list L1 are a subset of the members of L2, then (SUBSET L1 L2) is true. The function UNION produces the set union of L1 and L2. MAX-MEMBER returns the maximum number in LIST. MAKE-BALANCED-TREE constructs a balanced binary tree with N leaves. TREE-SIZE counts the number of leaves in a binary tree. TFIRSTN and TRESTN measure the number of leaves in the left-hand part of TREE to select elements out of LIST.

We have defined a number of small programs which we later use to compute values for our HDL primitives. These definitions appear in Figure 7.1. The definitions are not a part of the HDL we define later, but are used to help define the logical interpreter for our HDL primitives.

```
(B-BUF X)            = (IF X T F)
(B-NOT X)            = (NOT X)

(B-NAND  A B)        = (NOT (AND A B))
(B-NAND3 A B C)      = (NOT (AND A B C))
(B-NAND4 A B C D)    = (NOT (AND A B C D))

(B-OR   A B)         = (OR A B)
(B-OR3  A B C)       = (OR A B C)
(B-OR4  A B C D)     = (OR A B C D)

(B-EQUV X Y)         = (IF X (IF Y T F) (IF Y F T))
(B-XOR  X Y)         = (IF X (IF Y F T) (IF Y T F))

(B-AND  A B)         = (AND A B)
(B-AND3 A B C)       = (AND A B C)
(B-AND4 A B C D)     = (AND A B C D)

(B-NOR  A B)         = (NOT (OR A B))
(B-NOR3 A B C)       = (NOT (OR A B C))
(B-NOR4 A B C D)     = (NOT (OR A B C D))
```

Figure 7.1: Primitive Specifications

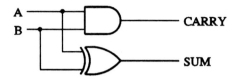

Figure 7.2: Half-Adder Circuit

7.4   Introduction to Circuit Generators

The purpose of circuit generator functions is to create parameterized libraries of circuits that can be generated when needed. We call our circuit constructor functions *generators* instead of *synthesizers* because we usually think of synthesizers as more fully exploring the design space than do our generator functions. That is not to say that our generator functions are conceptually any different than circuit synthesis functions; just that our generators may be simpler and are proven to be correct.

To give the flavor of our HDL formalization and the construction of verified circuit generator functions, we begin with the definition of an n-bit, ripple-carry adder generator. Later, we will consider the verification of this and several other circuit generator functions.

We represent circuits as a list of circuit boxes (modules). A well-formed circuit box contains four elements: a box name, a list of input names, a list of output names, and a circuit box body. We require the input and output names to be distinct. A circuit box body is just a set of wiring instructions interconnecting circuit boxes. The definition of our HDL includes simple Boolean gate as primitives. These gates are treated as pre-defined circuit boxes. A well-formed circuit box body does not admit wiring loops; i.e., only combinational logic without feedback is permitted. We think of circuit box input and output names as representing wire names.

Below is a circuit box constant for the half-adder whose schematic diagram is pictured in Figure 7.2.

```
'(HALF-ADDER (A B) (SUM CARRY) (((SUM)    (B-XOR A B))
                                ((CARRY) (B-AND A B))))
```

Circuit box **HALF-ADDER** has two inputs named **A** and **B** and two outputs named **SUM** and **CARRY**. The circuit box body is a list. Each circuit box body occurrence is composed of two elements: a list of outputs and a circuit box reference. Thus the first circuit box body occurrence is ((SUM) (B-XOR A B)); the output of circuit box reference (B-XOR A B) is connected to wire SUM.

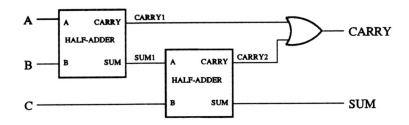

Figure 7.3: Full-Adder Circuit

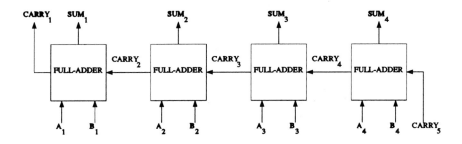

Figure 7.4: Four-bit, Ripple-carry Adder Circuit

A schematic for a full-adder is presented in Figure 7.3. The FULL-ADDER circuit box references the HALF-ADDER circuit box twice; thus to construct a full-adder circuit requires two half-adders.

```
'(FULL-ADDER (A B C) (SUM CARRY)
             (((SUM1 CARRY1) (HALF-ADDER A B))
              ((SUM CARRY2)  (HALF-ADDER SUM1 C))
              ((CARRY)       (B-OR CARRY1 CARRY2))))
```

We introduce the internal wires SUM1, CARRY1, and CARRY2 to interconnect the half-adders and the primitive B-OR gate.

One way to create an n-bit ripple-carry adder is to connect n full-adders together. For example, Figure 7.4 is a schematic diagram of a 4-bit ripple-carry adder, where A_1 is the most significant bit and the adder carries from right to left. The circuit box constant for this adder is below.

```
'(RIPPLE-ADDER₄
  (CARRY₅
   A₄ A₃ A₂ A₁
   B₄ B₃ B₂ B₁)
  (SUM₄ SUM₃ SUM₂ SUM₁ CARRY₁)
  (((SUM₄ CARRY₄) (FULL-ADDER A₄ B₄ CARRY₅))
   ((SUM₃ CARRY₃) (FULL-ADDER A₃ B₃ CARRY₄))
   ((SUM₂ CARRY₂) (FULL-ADDER A₂ B₂ CARRY₃))
   ((SUM₁ CARRY₁) (FULL-ADDER A₁ B₁ CARRY₂))))
```

More generally, we can define a function which creates a circuit box with an input variable that specifies the adder size. For instance, an n-bit circuit box might be written as follows.

```
'(RIPPLE-ADDERₙ
  (CARRYₙ₊₁
   Aₙ ... A₁
   Bₙ ... B₁)
  (SUMₙ ... SUM₁ CARRY₁)
  (((SUMₙ CARRYₙ) (FULL-ADDER Aₙ Bₙ CARRYₙ₊₁))
   ...
   ((SUM₁ CARRY₁) (FULL-ADDER A₁ B₁ CARRY₂))))
```

We construct our ripple-carry adder box generator in two parts: a top-level function which provides the box name, the input names, and the output names; and a function which creates the ripple-carry adder body. The function (V-ADDER-BODY N), just below, creates a list of N occurrences where each occurrence is a list of two elements: a list with two output names, SUM_i and $CARRY_i$; and a reference to a full-adder with parameters A_i, B_i, and $CARRY_{i+1}$. Our Boyer-Moore formulation of V-ADDER-BODY does not actually produce terms with subscripts as appears above; we simply abbreviate names of the form (LIT . i) as LIT_i.

```
(V-ADDER-BODY N)
  =
(IF (ZEROP N)
    NIL
    (CONS (LIST (LIST (CONS 'SUM N)
                      (CONS 'CARRY N))
                (LIST 'FULL-ADDER
                      (CONS 'A N)
                      (CONS 'B N)
                      (CONS 'CARRY (ADD1 N))))
          (V-ADDER-BODY (SUB1 N)))))
```

Below is the result of executing (V-ADDER-BODY 4). We use the symbol "==" to denote the result of evaluating a function.

```
(V-ADDER-BODY 4)
==
'((((SUM . 4) (CARRY . 4)) (FULL-ADDER (A . 4) (B . 4) (CARRY . 5)))
  (((SUM . 3) (CARRY . 3)) (FULL-ADDER (A . 3) (B . 3) (CARRY . 4)))
  (((SUM . 2) (CARRY . 2)) (FULL-ADDER (A . 2) (B . 2) (CARRY . 3)))
  (((SUM . 1) (CARRY . 1)) (FULL-ADDER (A . 1) (B . 1) (CARRY . 2))))
```

To complete our ripple-adder generator function, we now just need to put the V-ADDER-BODY into a circuit box with the appropriate input and output names; the function V-ADDER* produces such a circuit box. The function (GENERATE-NAMES LETTER N) produces a list of N names with LETTER as a seed.

```
(GENERATE-NAMES LETTER N)
=
(IF (ZEROP N)
    NIL
    (CONS (CONS LETTER N)
          (GENERATE-NAMES LETTER (SUB1 N)))))
```

```
(V-ADDER* N)
=
(LIST (CONS 'V-ADDER N)
      (CONS (CONS 'CARRY (ADD1 N))
            (APPEND (GENERATE-NAMES 'A N)
                    (GENERATE-NAMES 'B N)))
      (APPEND (GENERATE-NAMES 'SUM N)
              (LIST (CONS 'CARRY 1)))
      (V-ADDER-BODY N)))
```

Below is the result of evaluating (V-ADDER* 4).

```
(V-ADDER* 4)
==
'((V-ADDER . 4)                                        ; Name

  ((CARRY . 5)                                         ; Inputs
   (A . 4) (A . 3) (A . 2) (A . 1)
   (B . 4) (B . 3) (B . 2) (B . 1))

  ((SUM . 4) (SUM . 3) (SUM . 2) (SUM . 1)             ; Outputs
   (CARRY . 1))

  ;;Body
  ((((SUM . 4) (CARRY . 4)) (FULL-ADDER (A . 4) (B . 4) (CARRY . 5)))
   (((SUM . 3) (CARRY . 3)) (FULL-ADDER (A . 3) (B . 3) (CARRY . 4)))
   (((SUM . 2) (CARRY . 2)) (FULL-ADDER (A . 2) (B . 2) (CARRY . 3)))
   (((SUM . 1) (CARRY . 1)) (FULL-ADDER (A . 1) (B . 1) (CARRY . 2)))))
```

A complete circuit description is represented as a list of circuit boxes, which we call a *boxlist*. A complete boxlist for an *n*-bit, ripple-carry adder at the least contains a circuit box generated by (V-ADDER* *n*), along with an instance of FULL-ADDER and HALF-ADDER.

7.5   Boxlist Syntax

In this section we present formal specifications of well-formed circuit boxes and boxlists. We consider our simple HDL to be a formal abstraction of a combinational subset of a generic commercial CAD language. We believe that a boxlist that meets our syntactic criteria can be easily and mechanically translated into a commercial CAD language, and used as a reliable basis for a hardware realization of the formal model. We present an example of such a translator in Section 7.10. We have also defined a number of interpreters for our boxlists including interpreters for circuit values, fanout, and gate count. These interpreters are presented in Section 7.6. Since Boyer-Moore logic functions are total, these interpreters will compute a result for any boxlist; however, the interpretations are only meaningful for circuit descriptions that meet our syntactic requirements.

The following is an informal summary of the syntax requirements that our boxlist specification makes precise.

- Module names are unique to a boxlist.

- Modules are defined in terms of a small set of combinational primitives, or other modules defined in the boxlist. The boxlist contains a complete hierarchical description of every box.

- Input and output arities are consistent for each primitive and hierarchical reference.

- All nets are driven either by a module input or by exactly one primitive output. There are no busses.

- There is no feedback or potential feedback. In a top-to-bottom scan of the occurrence list for a module, all of the inputs nets for an occurrence must be either module inputs, or appear as outputs of previous occurrences.

The formal specification of a well-formed boxlist is given by BOXLIST-OKP. The definition of BOXLIST-OKP appears as Figure 7.5. BOXLIST-OKP checks that the boxlist as a whole has the correct structure, that boxes have unique names, and that the inputs and outputs names of each box are unique and disjoint.

```
(BOXLIST-OKP BOXLIST)
 =
(IF (NLISTP BOXLIST)
    (EQUAL BOXLIST NIL)

    (LET ((BOX (CAR BOXLIST))
          (REST (CDR BOXLIST)))
      (AND (LISTP BOX)
           (LISTP (CDR BOX))
           (LISTP (CDDR BOX))
           (LISTP (CDDDR BOX))
           (EQUAL (CDDDDR BOX) NIL)

           (LET ((NAME    (CAR BOX))
                 (INPUTS  (CADR BOX))
                 (OUTPUTS (CADDR BOX))
                 (BODY    (CADDDR BOX)))

             (AND (NOT (ASSOC NAME REST))
                  (PROPERP INPUTS)
                  (NOT (DUPLICATES? INPUTS))
                  (PROPERP OUTPUTS)
                  (NOT (DUPLICATES? OUTPUTS))
                  (DISJOINT INPUTS OUTPUTS)
                  (BODY-OKP BODY INPUTS OUTPUTS REST)
                  (BOXLIST-OKP REST)))))))
```

Figure 7.5: Definition of BOXLIST-OKP.

```
(BODY-OKP BODY SIGNALS OUTPUTS REST)
=
(IF (NLISTP BODY)
    ;; If the occurrence list is empty, then each output must
    ;; have been assigned.
    (AND (EQUAL BODY NIL)
         (SUBSET OUTPUTS SIGNALS))

    (LET ((OCCURRENCE (CAR BODY)))
      (AND
       (LISTP OCCURRENCE)
       (LISTP (CDR OCCURRENCE))
       (EQUAL (CDDR OCCURRENCE) NIL)

       (LET ((LHS (CAR OCCURRENCE))
             (RHS (CADR OCCURRENCE)))
         (AND
          (PROPERP LHS)
          (NOT (DUPLICATES? LHS))        ;These checks prohibit
          (DISJOINT LHS SIGNALS)          ;busses.
          (LISTP RHS)
          (PROPERP RHS)

          (LET ((MODULE-NAME (CAR RHS))
                (MODULE-ARGS (CDR RHS)))
            (LET ((PRIMP (PRIMP MODULE-NAME)))

              (AND
               (SUBSET MODULE-ARGS SIGNALS) ;Inputs must be driven.
               (IF PRIMP
                   ;; Arity checks for primitives.
                   (AND (EQUAL (CAR PRIMP) (LENGTH MODULE-ARGS))
                        (EQUAL (CDR PRIMP) (LENGTH LHS)))
                   ;; Existence and arity checks for submodules.
                   (LET ((SUBMODULE (ASSOC MODULE-NAME REST)))
                     (LET ((SUBMODULE-INPUTS  (CADR SUBMODULE))
                           (SUBMODULE-OUTPUTS (CADDR SUBMODULE)))
                       (AND SUBMODULE
                            (EQUAL (LENGTH MODULE-ARGS)
                                   (LENGTH SUBMODULE-INPUTS))
                            (EQUAL (LENGTH LHS)
                                   (LENGTH SUBMODULE-OUTPUTS))))))
               (BODY-OKP (CDR BODY) (APPEND LHS SIGNALS)
                         OUTPUTS REST)))))))))
```

Figure 7.6: Definition of BODY-OKP.

The syntax of the occurrence list of each box is specified by BODY-OKP. The definition of BODY-OKP appears as Figure 7.6. The BODY argument of BODY-OKP is the occurrence list; SIGNALS is initialized to the module input list by BOXLIST-OKP, and collects the internal signal names; OUTPUTS are the module output signals; and REST is the remainder of the boxlist, used to insure that all referenced submodules are defined.

BODY-OKP, as well as our interpretation functions defined in the next Section, refer to the specification PRIMP. PRIMP defines those names that are considered primitive, and returns a pair (*input-arity . output-arity*) for primitives, otherwise F for non-primitives.

```
(PRIMP FN)
  =
(CASE FN
      (B-BUF    (CONS 1 1)) (B-NOT   (CONS 1 1))
      (B-NAND   (CONS 2 1)) (B-NAND3 (CONS 3 1)) (B-NAND4 (CONS 4 1))
      (B-OR     (CONS 2 1)) (B-OR3   (CONS 3 1)) (B-OR4   (CONS 4 1))
      (B-EQUV   (CONS 2 1)) (B-XOR   (CONS 2 1))
      (B-AND    (CONS 2 1)) (B-AND3  (CONS 3 1)) (B-AND4  (CONS 4 1))
      (B-NOR    (CONS 2 1)) (B-NOR3  (CONS 3 1)) (B-NOR4  (CONS 4 1))
      (OTHERWISE F))
```

The choice of primitives is completely arbitrary; we have chosen a set of primitives that corresponds to the Boolean specification functions displayed in Figure 7.1.

7.6 Hardware Interpreters

To provide a meaning to our HDL circuit constants we have constructed four interpreters: a value interpreter, a delay interpreter, a loading interpreter, and an interpreter which counts the number of primitive references required by a circuit. Each of these interpreters share a similar structure, as each take well-formed circuits as input. Before giving the interpreter definitions, we consider the interpretation of the four-bit adder circuit presented earlier.

The value interpreter produces a logical value for a circuit reference given a boxlist and an association list containing values for signal names. BOXLIST-OKP only recognizes circuits which can be evaluated with a single depth first traversal. Each traversal terminates in the evaluation of a primitive. Consider the schematic representation of RIPPLE-ADDER$_4$ presented in Figure 7.7. To evaluate this circuit, each input must be bound to a value. To evaluate RIPPLE-ADDER$_4$, we must evaluate the four FULL-ADDER circuit boxes that comprise RIPPLE-ADDER$_4$. The evaluation begins with the least significant FULL-ADDER; the evaluation of FULL-ADDER proceeds in a manner similar to

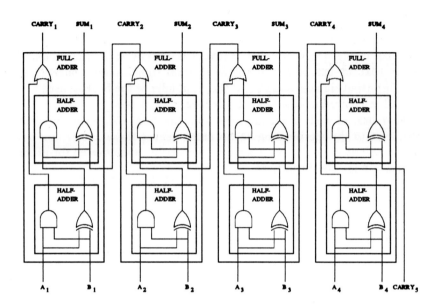

Figure 7.7: Exploded, Four-bit, Ripple-carry Adder Circuit

RIPPLE-ADDER₄. The evaluation continues until left with circuit boxes containing only primitives. The evaluation of primitives is defined by our circuit box evaluator.

7.6.1   The Logical Value Interpreter

Our logical value interpreter uses association lists to bind values to names. A value for **NAME** in the binding environment **ALIST** is defined by **EVAL-NAME**. **COLLECT-EVAL-NAME** uses **EVAL-NAME** to map a list of names to a list of values.

```
(EVAL-NAME NAME ALIST)
 =
(IF (NLISTP ALIST)
    F
    (IF (AND (LISTP (CAR ALIST))
             (EQUAL NAME (CAAR ALIST)))
        (BOOLFIX (CDAR ALIST))
        (EVAL-NAME NAME (CDR ALIST)))))
```

```
(COLLECT-EVAL-NAME ARGS ALIST)
=
(IF (NLISTP ARGS)
    NIL
    (CONS (EVAL-NAME (CAR ARGS) ALIST)
          (COLLECT-EVAL-NAME (CDR ARGS) ALIST)))
```

We give an example of the use of COLLECT-EVAL-NAME below.

```
(COLLECT-EVAL-NAME '(B C A) (LIST (CONS 'A F) (CONS 'B T)
                                  (CONS 'C T) (CONS 'D F)))
==
(LIST T T F)
```

The specification of evaluating primitive circuit boxes is given by the function HAPPLY. HAPPLY specifies a value for each function name recognized by PRIMP.

```
(HAPPLY FN ACTUALS)
=
(LET ((A (CAR ACTUALS))   (B (CADR ACTUALS))
      (C (CADDR ACTUALS)) (D (CADDDR ACTUALS)))
  (CASE FN
    (B-BUF   (LIST (B-BUF   A)))        (B-NOT   (LIST (B-NOT   A)))
    (B-NAND  (LIST (B-NAND  A B)))
    (B-NAND3 (LIST (B-NAND3 A B C)))    (B-NAND4 (LIST (B-NAND4 A B C D)))
    (B-OR    (LIST (B-OR    A B)))
    (B-OR3   (LIST (B-OR3   A B C)))    (B-OR4   (LIST (B-OR4   A B C D)))
    (B-EQUV  (LIST (B-EQUV  A B)))      (B-XOR   (LIST (B-XOR   A B)))
    (B-AND   (LIST (B-AND   A B)))
    (B-AND3  (LIST (B-AND3  A B C)))    (B-AND4  (LIST (B-AND4  A B C D)))
    (B-NOR   (LIST (B-NOR   A B)))
    (B-NOR3  (LIST (B-NOR3  A B C)))    (B-NOR4  (LIST (B-NOR4  A B C D)))
    (OTHERWISE F))))
```

The evaluation of a circuit box involves alternately evaluating circuit box references and evaluating circuit box bodies. The Boyer-Moore logic only allows definitions to refer to previously defined functions; therefore, we are unable to introduce two mutually recursive evaluation functions, but are forced to introduce a single function, HEVAL, containing both a circuit reference evaluator and a circuit box body evaluator. The definition of HEVAL appears as Figure 7.8. The control flag, FLAG, is used to specify the selection of an evaluator. If FLAG is not F, then FORM is assumed to be a circuit box reference and HEVAL returns a list of values, one value for each output of the referenced circuit box. If FLAG is F, the FORM is assumed to be a circuit box body and

```
(HEVAL FLAG FORM ALIST BOXLIST)
 =
(IF FLAG
    (LET ((FN   (CAR FORM))
          (ARGS (CDR FORM)))
      (LET ((ACTUALS (COLLECT-EVAL-NAME ARGS ALIST)))
        (IF (PRIMP FN)
            (HAPPLY FN ACTUALS)
            (LET ((BOX (ASSOC FN BOXLIST)))
              (IF BOX
                  (LET ((INPUTS  (CADR BOX))
                        (OUTPUTS (CADDR BOX))
                        (BODY    (CADDDR BOX)))
                    (COLLECT-EVAL-NAME
                     OUTPUTS
                     (HEVAL F BODY (PAIRLIST INPUTS ACTUALS)
                            (CDR BOXLIST))))
                  F)))))
    (IF (LISTP FORM)
        (LET ((OCCURRENCE (CAR FORM)))
          (LET ((LHS (CAR OCCURRENCE))
                (RHS (CADR OCCURRENCE)))
            (HEVAL F
                   (CDR FORM)
                   (APPEND (PAIRLIST LHS (HEVAL T RHS ALIST BOXLIST))
                           ALIST)
                   BOXLIST)))
        ALIST))
```

Figure 7.8: The Logical Value Interpreter, HEVAL.

HEVAL returns an association list that binds values for all outputs and internal signals.

If FLAG is not F and the occurrence is primitive, then HAPPLY is called. If FLAG is not F and the occurrence is not primitive, then the circuit box body is evaluated by HEVAL with an initial association list that binds the circuit box inputs to the actual parameters. If FLAG is F, HEVAL interprets FORM as a circuit box body; HEVAL performs the body evaluation recursively by evaluating each circuit box reference in a binding environment that contains the input bindings and output bindings for previously occurring circuit box references. After each reference in a circuit box body has been evaluated, an association list is returned containing bindings for every internal signal along with the original input bindings.

The termination of HEVAL is guaranteed by a decrease in a lexicographic measure of the arguments BOXLIST and FORM with each recursive call of HEVAL. In the call to HEVAL when FLAG is not F, the size of (CDR BOXLIST) is less than the size of BOXLIST. When FLAG is F, the size of FORM is decreasing in both recursive calls of HEVAL: the (CDR FORM) case is obvious, and in the other case HEVAL is called with FORM bound to RHS which is a subcomponent of OCCURRENCE, which itself is a subcomponent of FORM.

To demonstrate a HEVAL interpretation, we consider the evaluation of '(HALF-ADDER A B) in the following environment.

```
BOXLIST = (LIST '(HALF-ADDER (A B) (SUM CARRY)
                 (((SUM)   (B-XOR A B))
                 ((CARRY) (B-AND A B)))))

ALIST = (LIST (CONS 'X F) (CONS 'Y T))
```

Below is a partial trace of the evaluation of '(HALF-ADDER X Y), which finally simplifies to (LIST T F).

```
(HEVAL T '(HALF-ADDER X Y) ALIST BOXLIST)
==
(COLLECT-EVAL-NAME '(SUM CARRY)
              (HEVAL F '(((SUM) (B-XOR A B))
                        ((CARRY) (B-AND A B)))
                    (LIST (CONS 'A F) (CONS 'B T)) '()))
==

(COLLECT-EVAL-NAME '(SUM CARRY)
              (HEVAL F '(((CARRY) (B-AND A B)))
                    (LIST (CONS 'SUM T)
                        (CONS 'A F) (CONS 'B T)) '()))
==

(COLLECT-EVAL-NAME '(SUM CARRY)
              (HEVAL F '()
                    (LIST (CONS 'CARRY F) (CONS 'SUM T)
                        (CONS 'A F) (CONS 'B T))
                    '()))
==

(COLLECT-EVAL-NAME '(SUM CARRY)
              (LIST (CONS 'CARRY F) (CONS 'SUM T)
                  (CONS 'A F) (CONS 'B T)))
==
(LIST T F)
```

7.6.2   Other Interpreters

We have written several other interpreters: a delay interpreter which computes the delay for every output of any circuit box reference; a load interpreter which computes the input loading, output loading, and maximum internal loading of a circuit box reference; and an interpreter that counts the total number of primitives that must be evaluated for any circuit box reference.

GEVAL counts the number of primitives evaluated for any circuit box reference. For example, the ripple-carry adder circuit in Figure 7.7 contains 20 primitives gates. The structure of GEVAL is similar to HEVAL's structure. GEVAL's primitive apply function GAPPLY returns 1; however, it is possible to generalize GAPPLY to return different values for different primitives.

When given a circuit box reference, GEVAL calls GAPPLY if FORM is primitive; otherwise, GEVAL computes the number of primitives in the body of the current circuit box reference. Note, when the FLAG is not F, FORM is a circuit box reference, otherwise FORM is a circuit box body.

```
(GAPPLY FN) = 1

(GEVAL FLAG FORM BOXLIST)
  =
(IF FLAG
    (LET ((FN   (CAR FORM))
          (ARGS (CDR FORM)))
      (IF (PRIMP FN)
          (GAPPLY FN)
          (LET ((BOX (ASSOC FN BOXLIST)))
            (IF BOX
                (LET ((BODY (CADDDR BOX)))
                  (GEVAL F BODY (CDR BOXLIST)))
                F))))

    (IF (LISTP FORM)
        (LET ((OCCURRENCE (CAR FORM)))
          (LET ((RHS (CADR OCCURRENCE)))
            (PLUS (GEVAL T RHS BOXLIST)
                  (GEVAL F (CDR FORM) BOXLIST))))
        0))
```

DEVAL computes the maximum delay for every output from every relevant input. Unique initial delays may be provided for every input. In DAPPLY we have chosen the delay for each primitive to be one more than the maximum delay of its inputs. We could define DAPPLY to reflect delays for a particular implementation technology.

```
(DAPPLY FN ACTUALS)
 =
(LIST (ADD1 (MAX-MEMBER ACTUALS))))
```

DEVAL operates in a manner just like HEVAL, but instead of computing the value for every output, it computes the delay for every output. The initial ALIST contains the initial delays for every input.

```
(DEVAL FLAG FORM ALIST BOXLIST)
 =
(IF FLAG
    (LET ((FN   (CAR FORM))
          (ARGS (CDR FORM)))
      (LET ((DELAYS (COLLECT-ASSOC ARGS ALIST)))
        (IF (PRIMP FN)
            (DAPPLY FN DELAYS)
            (LET ((BOX (ASSOC FN BOXLIST)))
              (IF BOX
                  (LET ((INPUTS  (CADR BOX))
                        (OUTPUTS (CADDR BOX))
                        (BODY    (CADDDR BOX)))
                    (COLLECT-ASSOC
                     OUTPUTS
                     (DEVAL F BODY (PAIRLIST INPUTS DELAYS)
                            (CDR BOXLIST))))
                  F)))))
    (IF (LISTP FORM)
        (LET ((OCCURRENCE (CAR FORM)))
          (LET ((LHS (CAR OCCURRENCE))
                (RHS (CADR OCCURRENCE)))
            (DEVAL F
                   (CDR FORM)
                   (APPEND (PAIRLIST LHS
                                     (DEVAL T RHS ALIST BOXLIST))
                           ALIST)
                   BOXLIST)))
        ALIST))
```

The loading interpreter, LEVAL, returns three items: the input loadings, the output loadings, and the maximum internal loading. Each primitive input has a loading of one. Primitive output loadings are zero; however, the output loadings of a circuit box are not always zero, as box outputs may be wired to other internal inputs. The loadings for the primitive circuit boxes are specified by LAPPLY. The LEVAL function concerns itself with the structure of a circuit box reference, not about values it may compute. Loadings are returned as a list with the input loading first, the output loading second, and the maximum internal loading last.

```
(LAPPLY FN)
 =
(LIST (MAKE-LIST (CAR (PRIMP FN)) 1) (LIST 0) 1))
```

ALIST is used in LEVAL to supply initial loadings for every input.

```
(LEVAL FLAG FORM ALIST BOXLIST)
 =
(IF FLAG
    (LET ((FN   (CAR FORM))
          (ARGS (CDR FORM)))
      (IF (PRIMP FN)
          (LAPPLY FN)
          (LET ((BOX (ASSOC FN BOXLIST)))
            (IF BOX
                (LET ((INPUTS  (CADR BOX))
                      (OUTPUTS (CADDR BOX))
                      (BODY    (CADDDR BOX)))
                  (LET ((LEVAL (LEVAL F BODY (PAIRLIST INPUTS 0)
                                      (CDR BOXLIST))))
                    (LET ((ALIST (CAR LEVAL))
                          (NEW-MAX-FAN (CDR LEVAL)))
                      (LIST (COLLECT-ASSOC INPUTS ALIST)
                            (COLLECT-ASSOC OUTPUTS ALIST)
                            NEW-MAX-FAN))))
                F))))
    (IF (LISTP FORM)
        (LET ((OCCURRENCE (CAR FORM)))
          (LET ((LHS (CAR OCCURRENCE))
                (RHS (CADR OCCURRENCE)))
            (LET ((TRIPLE (LEVAL T RHS NIL BOXLIST)))
              (LET ((I-COSTS (CAR TRIPLE))
                    (O-COSTS (CADR TRIPLE))
                    (TERM-MAX-COST (CADDR TRIPLE)))
                (LET ((NEW-ALIST
                        (APPEND (PAIRLIST LHS O-COSTS)
                                (ADD-TO-ALIST (CDR RHS)
                                              I-COSTS ALIST))))
                  (LET ((RETURN-ALIST
                          (CAR (LEVAL F (CDR FORM) NEW-ALIST
                                     BOXLIST)))
                        (RETURN-MAX-COST
                          (CDR (LEVAL F (CDR FORM) NEW-ALIST
                                     BOXLIST))))
                    (CONS RETURN-ALIST
                          (MAX TERM-MAX-COST RETURN-MAX-COST))))))))
        (CONS ALIST (MAX-ALIST-VAL ALIST))))
```

It is easy to imagine the definition of other interpreters. For instance, a critical path interpreter could be written, delays could be computed with fanout information taken into account, or power requirements could be estimated.

7.7   A Simple Circuit Generator

A consequence of using constants to specify hardware circuits is an ability to verify functions which generate circuit descriptions. To demonstrate verifying circuit generators, we consider the proof of the ripple-carry adder generator presented earlier. The ripple-carry adder example is often used in descriptions of formal hardware verification methodologies; thus the interested reader can directly compare this treatment with other approaches [15, 9, 19, 16]. We are not aware of other work specifically attempting to verify circuit generator functions.

It is important to clarify exactly what we intend to verify. We can only prove the correctness of the interpretation of a circuit box reference with respect to a boxlist that completely defines the hierarchical structure of the circuit. The verification of circuit generators is three part: we specify a circuit box generator, we specify a predicate which recognizes a boxlist that contains instances of the circuit box generator, and finally we prove the correctness of a reference to the circuit box with respect to a specification function, assuming our predicate holds for the boxlist. The structure of these predicates mimics the hierarchy of the design, as do the structure of the proofs. We present the verification of our ripple-carry adder generator in a bottom-up manner.

A lemma stating the correctness of a reference to an exclusive OR primitive is shown below. In the case of primitives, like exclusive OR, we need not construct circuit generators, and the predicates which recognize primitives in a boxlist are always true.

```
(B-XOR& BOXLIST) = T

(IMPLIES (B-XOR& BOXLIST)
         (EQUAL (HEVAL T (LIST 'B-XOR X Y) ALIST BOXLIST)
                (LET ((X (EVAL-NAME X ALIST))
                      (Y (EVAL-NAME Y ALIST)))
                  (LIST (B-XOR X Y)))))
```

This lemma indicates that the evaluation with HEVAL of any reference of the form (LIST 'B-XOR X Y) is precisely the same as applying the specification function B-XOR to the values of X and Y in ALIST. We now take as given that the boxlist recognizers for our primitives are true, and that we have proven the correctness of the evaluation of each primitive with respect to their specification functions.

The verification of a circuit box generator for a half-adder proceeds in three parts: we define the circuit box generator HALF-ADDER*, we define a predicate, HALF-ADDER&, that recognizes boxlists containing an instance of our half-adder circuit box, and finally we prove the correctness of a reference to a half-adder.

```
(HALF-ADDER*)
  =
'(HALF-ADDER (A B) (SUM CARRY)
             (((SUM)   (B-XOR A B))
              ((CARRY) (B-AND A B))))

(HALF-ADDER& BOXLIST)
  =
(AND (EQUAL (ASSOC 'HALF-ADDER BOXLIST) (HALF-ADDER*))
     (B-XOR& (CDR BOXLIST))
     (B-AND& (CDR BOXLIST)))

(IMPLIES (HALF-ADDER& BOXLIST)
         (EQUAL (HEVAL T (LIST 'HALF-ADDER A B) ALIST BOXLIST)
                (LET ((A (EVAL-NAME A ALIST))
                      (B (EVAL-NAME B ALIST)))
                  (LIST (B-XOR A B)
                        (B-AND A B)))))
```

We verify the correctness of our full-adder in the same manner as we verified our half-adder. We first introduce the specification functions FULL-ADDER-SUM and FULL-ADDER-CARRY for our full-adder.

```
(FULL-ADDER-SUM A B C)
  =
(IF A
    (IF B (IF C T F) (IF C F T))
    (IF B (IF C F T) (IF C T F)))

(FULL-ADDER-CARRY A B C)
  =
(OR (AND A (OR B C)) (AND B C))
```

Shown below is our full-adder box generator, FULL-ADDER*, a predicate which recognizes boxlists that contain an instance of FULL-ADDER*, and a lemma demonstrating the correctness of evaluating a reference to a full-adder.

```
(FULL-ADDER*)
  =
'(FULL-ADDER (A B C) (SUM CARRY)
             (((SUM1 CARRY1) (HALF-ADDER A B))
              ((SUM CARRY2) (HALF-ADDER SUM1 C))
              ((CARRY) (B-OR CARRY1 CARRY2))))
```

```
(FULL-ADDER& BOXLIST)
 =
(AND (EQUAL (ASSOC 'FULL-ADDER BOXLIST) (FULL-ADDER*))
     (HALF-ADDER& (CDR BOXLIST))
     (B-OR& (CDR BOXLIST)))

(IMPLIES (FULL-ADDER& BOXLIST)
         (EQUAL (HEVAL T (LIST 'FULL-ADDER A B C) ALIST BOXLIST)
                (LET ((A (EVAL-NAME A ALIST))
                      (B (EVAL-NAME B ALIST))
                      (C (EVAL-NAME C ALIST)))
                  (LIST (FULL-ADDER-SUM A B C)
                        (FULL-ADDER-CARRY A B C)))))
```

We are now ready to verify the correctness of the ripple-carry adder generator with respect to our Boolean addition specification V-ADDER.

```
(V-ADDER C A B)
 =
(IF (NLISTP A)
    (CONS (IF C T F) NIL)
    (CONS (XOR C (XOR (CAR A) (CAR B)))
          (V-ADDER (OR (AND (CAR A) (CAR B))
                       (OR (AND (CAR A) C)
                           (AND (CAR B) C)))
                   (CDR A)
                   (CDR B))))
```

V-ADDER specifies Boolean addition of bit-vectors A and B, with an initial input carry C.

Our ripple-carry adder generator, V-ADDER*, generates a circuit box whose body contains N references to full-adders; these references are constructed by the helper function V-ADDER-BODY.

```
(V-ADDER-BODY N)
 =
(IF (ZEROP N)
    NIL
    (CONS (LIST (LIST (CONS 'SUM N) (CONS 'CARRY N))
                (LIST 'FULL-ADDER (CONS 'A N) (CONS 'B N)
                      (CONS 'CARRY (ADD1 N))))
          (V-ADDER-BODY (SUB1 N))))
```

```
(V-ADDER* N)
  =
(LIST (CONS 'V-ADDER N)                              ; Name
      (CONS (CONS 'CARRY (ADD1 N))                   ; Inputs
            (APPEND (GENERATE-NAMES 'A N)
                    (GENERATE-NAMES 'B N)))
      (APPEND (GENERATE-NAMES 'SUM N)                ; Outputs
              (LIST (CONS 'CARRY 1)))
      (V-ADDER-BODY N))                              ; Body
```

We verify the correctness of our helper function V-ADDER-BODY before attempting to verify V-ADDER*. The lemma below shows that the association list returned by HEVAL from the evaluation of the body of the adder binds the output names correctly.

```
(IMPLIES
  (AND (FULL-ADDER& BOXLIST)
       (EQUAL C (EVAL-NAME (CONS 'CARRY (ADD1 N)) ALIST))
       (EQUAL A (COLLECT-EVAL-NAME (GENERATE-NAMES 'A N) ALIST))
       (EQUAL B (COLLECT-EVAL-NAME (GENERATE-NAMES 'B N) ALIST)))
  (EQUAL (COLLECT-EVAL-NAME (APPEND (GENERATE-NAMES 'SUM N)
                                    (LIST (CONS 'CARRY 1)))
                           (HEVAL F
                                  (V-ADDER-BODY N)
                                  ALIST
                                  BOXLIST))
         (V-ADDER C A B)))
```

We prove that the HEVAL interpretation of a reference to a V-ADDER circuit box is equal to Boolean addition only if the BOXLIST contains an instance of V-ADDER* (the adder generator), and instances of the HALF-ADDER and FULL-ADDER circuits. With the interpretation of the adder body in hand, it is a simple matter to complete the proof of adder circuit references.

```
(V-ADDER& BOXLIST N)
  =
(AND (EQUAL (ASSOC (CONS 'V-ADDER N) BOXLIST) (V-ADDER* N))
     (FULL-ADDER& (CDR BOXLIST)))
```

```
(IMPLIES (AND (V-ADDER& BOXLIST N)
              (EQUAL (LENGTH A) N)
              (EQUAL (LENGTH B) N))
         (EQUAL (HEVAL T (CONS (CONS 'V-ADDER N)
                               (CONS C (APPEND A B)))
                       ALIST BOXLIST)
                (LET ((C (EVAL-NAME C ALIST))
                      (A (COLLECT-EVAL-NAME A ALIST))
                      (B (COLLECT-EVAL-NAME B ALIST)))
                  (V-ADDER C A B))))
```

We have not demonstrated a well-formed boxlist that satisfies V-ADDER&. V-ADDER-BOXLIST generates a boxlist that is both well-formed and satisfies V-ADDER&.

```
(V-ADDER-BOXLIST N) = (LIST (V-ADDER* N)
                            (FULL-ADDER*)
                            (HALF-ADDER*))
```

V-ADDER-BOXLIST generates a boxlist with three circuit boxes: an n-bit ripple-carry adder, a full-adder, and a half-adder. To prove that V-ADDER-BOXLIST satisfies V-ADDER& is trivial. It is also possible to prove that V-ADDER-BOXLIST generates a well-formed boxlist; however, we do not prove either property. In general, we never prove that our parameterized circuit generators are well-formed or that they satisfy any particular circuit boxlist predicate. Instead, we prove that instances of boxlists created by parameterized boxlist generators are well-formed by executing BOXLIST-OKP. Likewise, we show that instances of circuit boxlists satisfy their circuit boxlist predicates by executing these predicates on boxlists of interest. For example, when we generate a boxlist with V-ADDER-BOXLIST we check that its result is complete and well-formed by executing the functions V-ADDER& and BOXLIST-OKP.

We can extend the lemma above by observing that V-ADDER can be used to add natural numbers when they are represented as bit-vectors. We define the abstraction function V-TO-NAT that converts a bit-vector into a natural number.

```
(V-TO-NAT X) = (IF (NLISTP X)
                   0
                   (PLUS (IF (CAR X) 1 0)
                         (TIMES 2 (V-TO-NAT (CDR X))))))
```

We are then able to prove the lemma below, which shows that V-ADDER specifies addition of bit-vector representations of natural numbers.

```
(IMPLIES (AND (BVP A)
              (BVP B)
              (EQUAL (LENGTH A) (LENGTH B)))
         (EQUAL (V-TO-NAT (V-ADDER C A B))
                (PLUS (IF C 1 0)
                      (V-TO-NAT A)
                      (V-TO-NAT B))))
```

With this lemma, we are able to finally prove the lemma below.

```
(IMPLIES (AND (V-ADDER& BOXLIST N)
              (EQUAL (LENGTH A) N)
              (EQUAL (LENGTH B) N))
         (EQUAL (V-TO-NAT (HEVAL T (CONS (CONS 'V-ADDER N)
                                         (CONS C (APPEND A B)))
                                 ALIST BOXLIST))
                (LET ((C (EVAL-NAME C ALIST))
                      (A (COLLECT-EVAL-NAME A ALIST))
                      (B (COLLECT-EVAL-NAME B ALIST)))
                     (PLUS (IF C 1 0)
                           (V-TO-NAT A)
                           (V-TO-NAT B)))))
```

The presentation above sets the theme for the next two sections. In the next section we employ heuristics to generate an adder with a minimum cost. Afterward, we outline an ALU generator.

7.8   A Heuristically Guided Adder Generator

The use of programs to generate provably correct hardware circuits provides great freedom in selecting designs through the use of heuristic guidance. Here we outline an adder circuit generator that selects a ripple-carry or propagate-generate adder based on a comparison of their costs.

The function COST computes the cost of a circuit reference by adding the maximum circuit delay to the result of dividing the number of primitive gates required to build the circuit by three. To aid in the computation of the cost we use DEVAL to calculate circuit delays and GEVAL to compute the number of required gates.

```
(COST FORM BOXLIST)
  =
(PLUS (QUOTIENT (GEVAL T FORM BOXLIST) 3)
      (MAX-MEMBER (DEVAL T FORM (PAIRLIST (CDR FORM) 0) BOXLIST)))
```

We use the circuit box generated by ADDER* to provide an indirect reference to either a ripple-carry adder or a propagate-generate adder. The particular reference returned by ADDER* is chosen by a comparison of the cost for an n-bit ripple-carry adder with the cost for an n-bit propagate-generate adder. To make this comparison, ADDER* actually creates instances of both the ripple-carry adder and the propagate-generate adder, measures their costs, and returns a reference to the adder with the least cost. Propagate-generate adders are characterized by their tree-structured look-ahead logic. To specify a propagate-generate adder of n-bits, we construct a balanced binary tree with n leaves; the structure of this tree specifies the organization of the look-ahead logic.

The type of adder for which an indirect reference is created is specified in the last three lines of the definition of ADDER*. The remainder of the ADDER* definition is used to make the comparison between the two adder types.

```
(ADDER* N)
=
(LET ((TREE (MAKE-BALANCED-TREE N)))
  (LET ((RIPPLE-NAME (CONS 'V-ADDER N))
        (PG-NAME     (CONS 'TV-ADDER TREE))
        (ARGS        (CONS (CONS 'CARRY (ADD1 N))
                           (APPEND (GENERATE-NAMES 'A N)
                                   (GENERATE-NAMES 'B N))))
        (OUTPUTS     (GENERATE-NAMES 'OUT (ADD1 N))))
    (LET ((RIPPLE-BOXLIST (V-ADDER-BOXLIST N))
          (PG-BOXLIST     (TV-ADDER-BOXLIST TREE))
          (RIPPLE-FORM    (CONS RIPPLE-NAME ARGS))
          (PG-FORM        (CONS PG-NAME ARGS)))
      (LET ((RIPPLE-COST (COST RIPPLE-FORM RIPPLE-BOXLIST))
            (PG-COST     (COST PG-FORM PG-BOXLIST)))
        (LIST (CONS 'ADDER N)                          ; Name
              ARGS                                     ; Inputs
              OUTPUTS                                  ; Outputs
              (IF (LESSP RIPPLE-COST PG-COST)          ; Body
                  (LIST (LIST OUTPUTS RIPPLE-FORM))
                  (LIST (LIST OUTPUTS PG-FORM))))))))
```

For example, an instance of ADDER* with N equals four is shown below.

```
(ADDER* 4)
 ==
'((ADDER . 4)
  ((CARRY . 5)
   (A . 4) (A . 3) (A . 2) (A . 1)
   (B . 4) (B . 3) (B . 2) (B . 1))
  ((OUT . 5) (OUT . 4) (OUT . 3) (OUT . 2) (OUT . 1))
  ((((OUT . 5) (OUT . 4) (OUT . 3) (OUT . 2) (OUT . 1))
   ((V-ADDER . 4)
    (CARRY . 5)
    (A . 4) (A . 3) (A . 2) (A . 1)
    (B . 4) (B . 3) (B . 2) (B . 1)))))
```

We prove a lemma that demonstrates the correctness of our adder generator given any boxlist that satisfies ADDER&. TV-ADDER& is a predicate that recognizes boxlists with instances of propagate-generate adders.

```
(ADDER& BOXLIST N)
 =
(AND (EQUAL (ASSOC (CONS 'ADDER N) BOXLIST) (ADDER* N))
     (LET ((TREE (MAKE-BALANCED-TREE N)))
       (AND (V-ADDER& (CDR BOXLIST) N)
            (TV-ADDER& (CDR BOXLIST) TREE))))

(IMPLIES (AND (ADDER& BOXLIST N)
              (EQUAL (LENGTH A) N)
              (EQUAL (LENGTH B) N)
              (NOT (ZEROP N)))
         (EQUAL (HEVAL T (CONS (CONS 'ADDER N)
                               (CONS C (APPEND A B)))
                       ALIST BOXLIST)
                (V-ADDER (EVAL-NAME C ALIST)
                         (COLLECT-EVAL-NAME A ALIST)
                         (COLLECT-EVAL-NAME B ALIST))))
```

This lemma has the same form as the lemma we previously proved for our ripple-carry adder. For simplicity, we impose the condition in ADDER& that BOXLIST contains both ripple-carry and propagate-generate adder definitions.

We have computed the costs for different sized ripple-carry and propagate-generate adders and provide a comparison of these costs in Table 7.1. The maximum delay and number of gates for a ripple-carry adder increase linearly with the size of the adder. The propagate-generate maximum delay increases logarithmically with the size of the adder, while the size increases at a roughly linear rate. Notice that an adder larger than 26 bits must be generated before the propagate-generate adder becomes cheaper than the ripple-carry adder.

Adder Size	Ripple Gate Count	Ripple Max Delay	Ripple Cost	Prop-Gen Gate Count	Prop-Gen Max Delay	Prop-Gen Cost
1 bit	5	3	4	6	3	5
2 bits	10	5	8	15	5	10
4 bits	20	9	15	33	8	19
8 bits	40	17	30	69	12	35
16 bits	80	33	59	132	15	59
25 bits	125	51	92	222	19	93
26 bits	130	53	96	231	19	96
27 bits	135	55	100	240	19	99
32 bits	160	65	118	285	20	115
64 bits	320	129	235	573	24	215
128 bits	640	257	470	1149	28	411

Table 7.1: Ripple-Carry, Generate-Propagate Adder Cost Comparison

The point to this section is to demonstrate how heuristics can be used to control the generation of provably correct circuits. Our adder generator selects an "appropriate" adder based on our cost function. Note that any other scheme could have been substituted to select which adder is specified.

7.9 An ALU Generator

We have verified a circuit box generator function for an n-bit ALU [1]. There are two parts to the ALU verification: the proof that the top-level Boolean ALU specification implements mathematical functions and the proof that the HDL circuit generated implements the top-level Boolean ALU specification. Here we present the top-level Boolean ALU specification, a lemma stating the correctness of the ALU design, and sketch of a part of the internal implementation of the ALU.[4]

---

[4] We have proved that the top-level Boolean ALU specification implements a number of Boolean, natural number, and integer operations. Each ALU operation has been verified to correctly implement one or more functions. For instance, if a logical OR operation is selected, then we have proved that our Boolean ALU specification computes the logical OR; however, if an arithmetic shift right is selected we have proved not only the logical properties of this operation, but we have also proved that this operation implements a division by two for a bit-vector representing a natural number or an integer. For addition operations we have proved the correctness of the Boolean ALU specification with respect to natural number and integer addition; likewise for subtraction. Here, we give the ALU specification as a Boolean function, as this is what the actual ALU hardware should compute.

OP-CODE	Result	Description
0000	a	Move
0001	$a + 1$	Increment
0010	$b + a + c$	Add with carry
0011	$b + a$	Add
0100	$0 - a$	Negation
0101	$a - 1$	Decrement
0110	$b - a - c$	Subtract with borrow
0111	$b - a$	Subtract
1000	$a \gg 1$	Rotate right, shifted through carry
1001	$a \gg 1$	Arithmetic shift right, top bit duplicated
1010	$a \gg 1$	Logical shift right, top bit zero
1011	$b \veebar a$	Exclusive Or
1100	$b \vee a$	Or
1101	$b \wedge a$	And
1110	$\neg a$	Not
1111	a	Move

Table 7.2: Informal ALU Operation Summary

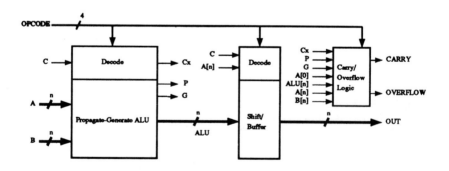

Figure 7.9: Internal Organization of the ALU

Our Boolean ALU specification, V-ALU, plays the same role that V-ADDER played in the verification of our ripple-carry adder. V-ALU is the specification that our ALU circuit generator is verified to implement. An informal summary of V-ALU is presented in Table 7.2.

V-ALU is the top-level Boolean specification for our ALU. It requires four inputs: C, a Boolean carry input; A and B, bit-vector inputs; and a four-bit op-code, OP. V-ALU returns a bit-vector that is two bits longer than bit-vector A: the first bit is the carry output, the second bit is the overflow output, and the remainder is the result bit-vector.

```
(V-ALU C A B OP)
=
(COND ((EQUAL OP (BC 'B0000)) (CVBV F F (V-BUF A)))
      ((EQUAL OP (BC 'B0001)) (CVBV-INC A))
      ((EQUAL OP (BC 'B0010)) (CVBV-V-ADDER C A B))
      ((EQUAL OP (BC 'B0011)) (CVBV-V-ADDER F A B))
      ((EQUAL OP (BC 'B0100)) (CVBV-NEG A))
      ((EQUAL OP (BC 'B0101)) (CVBV-DEC A))
      ((EQUAL OP (BC 'B0110)) (CVBV-V-SUBTRACTER C A B))
      ((EQUAL OP (BC 'B0111)) (CVBV-V-SUBTRACTER F A B))
      ((EQUAL OP (BC 'B1000)) (CVBV-V-ROR C A))
      ((EQUAL OP (BC 'B1001)) (CVBV (BITN 0 A) F (V-ASR A)))
      ((EQUAL OP (BC 'B1010)) (CVBV (BITN 0 A) F (V-LSR A)))
      ((EQUAL OP (BC 'B1011)) (CVBV F F (V-XOR A B)))
      ((EQUAL OP (BC 'B1100)) (CVBV F F (V-OR  A B)))
      ((EQUAL OP (BC 'B1101)) (CVBV F F (V-AND A B)))
      ((EQUAL OP (BC 'B1110)) (CVBV F F (V-NOT A)))
      (T                      (CVBV F F (V-BUF A))))
```

Of the many different possible implementations for the V-ALU specification, we have verified an ALU implementation which includes a propagate-generate look-ahead scheme for additions and subtractions. Previously, we have verified implementations of V-ALU with a ripple-carry approach [19] and a bit-slice approach [20]; however, these previous approaches did not specify their implementations with an HDL but required a program to extract a netlist from Boyer-Moore definitions which represented the design specification. Our ALU implementation is produced by a generator, like V-ADDER*, but it generates quite a large number of boxes with fairly complicated interconnections.

A block diagram of the internal structure of our ALU implementation is shown in Figure 7.9. The ALU is divided into three main parts: a propagate-generate ALU, which performs logical and arithmetic operations; a shift/buffer unit, which performs the right shift operations; and a carry/overflow unit which computes carry and overflow.

Our boxlist recognizer for our ALU is NEW-ALU&, which is presented below. We do not present the individual predicates that NEW-ALU& references.

```
(NEW-ALU& BOXLIST TREE)
 =
(AND (EQUAL (ASSOC (CONS 'NEW-ALU TREE) BOXLIST)
            (NEW-ALU* TREE))
     (B-BUF&             (CDR BOXLIST))
     (MPG&               (CDR BOXLIST))
     (TV-SHIFT-OR-BUF&   (CDR BOXLIST) TREE)
     (CARRY-IN-HELP&     (CDR BOXLIST))
     (TV-ALU-HELP&       (CDR BOXLIST) TREE)
     (T-CARRY&           (CDR BOXLIST))
     (CARRY-OUT-HELP&    (CDR BOXLIST))
     (OVERFLOW-HELP&     (CDR BOXLIST)))
```

The lemma stating the correctness of our ALU is below.

```
(IMPLIES
 (AND (NEW-ALU& BOXLIST TREE)
      (EQUAL (LENGTH A) (TREE-SIZE TREE))
      (EQUAL (LENGTH B) (TREE-SIZE TREE)))
 (EQUAL (HEVAL T (CONS (CONS 'NEW-ALU TREE)
                       (CONS C
                             (APPEND A
                                     (APPEND B
                                             (LIST OP0 OP1 OP2 OP3)))))
              ALIST BOXLIST)
        (V-ALU (EVAL-NAME C ALIST)
               (COLLECT-EVAL-NAME A ALIST)
               (COLLECT-EVAL-NAME B ALIST)
               (LIST (EVAL-NAME OP0 ALIST)
                     (EVAL-NAME OP1 ALIST)
                     (EVAL-NAME OP2 ALIST)
                     (EVAL-NAME OP3 ALIST)))))
```

That is, when the predicate NEW-ALU& is satisfied, then the HEVAL of NEW-ALU$_{TREE}$ satisfies our V-ALU specification function.

Instead of presenting the entire specification and implementation of our ALU generator, we have chosen to present one particular part of the ALU. The most interesting part of our ALU generator is the propagate-generate arithmetic/logical unit (PG-ALU). The PG-ALU performs all of the ALU operations expect for the shift operations and the computation of the carry and overflow outputs. The PG-ALU circuit has 4 inputs; bit-vector data inputs A and B, a carry input C, and an 8-bit control vector MPG.

The PG-ALU circuit generator requires a single parameter, TREE. The number of leaves in the tree defines the size of the data input vectors, and the structure of the tree determines the configuration of the carry look-ahead logic. Each leaf of the tree is represented in the final circuit by an instance

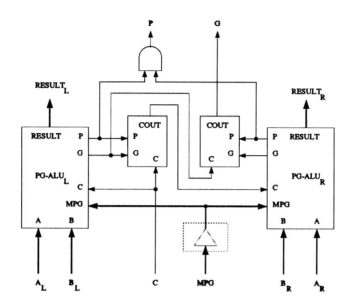

Figure 7.10: PG-ALU Circuit Organization

of an ALU function cell, T-CELL. Each internal node of the tree is represented in the final circuit by two PG-ALU modules connected together with carry look-ahead logic.

The definition of the PG-ALU generator TV-ALU-HELP* is below.

```
(TV-ALU-HELP* TREE)
=
(LET ((A-NAMES (GENERATE-NAMES 'A (TREE-SIZE TREE)))
      (B-NAMES (GENERATE-NAMES 'B (TREE-SIZE TREE)))
      (OUT-NAMES (GENERATE-NAMES 'OUT (TREE-SIZE TREE)))
      (MPGNAMES (GENERATE-NAMES 'MPG 8)))
  (LIST (CONS 'TV-ALU-HELP TREE)          ;; Name
        (CONS 'C                          ;; Inputs
              (APPEND A-NAMES
                      (APPEND B-NAMES MPGNAMES)))
        (CONS 'P (CONS 'G OUT-NAMES))     ;; Outputs
        (TV-ALU-HELP-BODY TREE)))         ;; Body
```

The body of the PG-ALU is created by TV-ALU-HELP-BODY, which appears in Figure 7.11. The schematic diagram in Figure 7.10 shows the interconnections generated by TV-ALU-HELP-BODY for each internal tree node. Notice that TV-ALU-HELP-BODY buffers the control lines only if a control line may drive more than 8 primitives. Also notice that TV-ALU-HELP-BODY is not a recursive func-

```
(TV-ALU-HELP-BODY (TREE)
=
(LET ((A-NAMES (GENERATE-NAMES 'A (TREE-SIZE TREE)))
      (B-NAMES (GENERATE-NAMES 'B (TREE-SIZE TREE)))
      (OUT-NAMES (GENERATE-NAMES 'OUT (TREE-SIZE TREE)))
      (MPGNAMES (GENERATE-NAMES 'MPG 8))
      (MPGNAMES* (GENERATE-NAMES 'MPG* 8)))
 (LET ((LEFT-A-NAMES (TFIRSTN A-NAMES TREE))
      (RIGHT-A-NAMES (TRESTN A-NAMES TREE))
      (LEFT-B-NAMES (TFIRSTN B-NAMES TREE))
      (RIGHT-B-NAMES (TRESTN B-NAMES TREE))
      (LEFT-OUT-NAMES (TFIRSTN OUT-NAMES TREE))
      (RIGHT-OUT-NAMES (TRESTN OUT-NAMES TREE)))
  (LET ((BUFFER? (EQUAL (REMAINDER (SUB1 (TREE-HEIGHT TREE)) 3) 0)))
   (LET ((MPGNAMES? (IF BUFFER? MPGNAMES* MPGNAMES)))
    (IF (NLISTP TREE)
        (LIST (LIST '(P G (OUT . 1))
                    (CONS 'T-CELL (CONS 'C
                                        (CONS '(A . 1)
                                              (CONS '(B . 1)
                                                    MPGNAMES))))))
      ;; We buffer MPG if more than 8 loads may be driven.
      (APPEND
       (IF BUFFER?
           (LIST (LIST MPGNAMES* (CONS (CONS 'V-BUF 8) MPGNAMES)))
           NIL)
       (LIST
        (LIST                           ;; The left-hand-side ALU.
         (CONS 'PL (CONS 'GL LEFT-OUT-NAMES))
         (CONS (CONS 'TV-ALU-HELP (CAR TREE))
               (CONS 'C (APPEND LEFT-A-NAMES
                                (APPEND LEFT-B-NAMES MPGNAMES?)))))
        '((CL) (T-CARRY C PL GL))       ;; The intermediate carry.
        (LIST                           ;; The right-hand-side ALU.
         (CONS 'PR (CONS 'GR RIGHT-OUT-NAMES))
         (CONS (CONS 'TV-ALU-HELP (CDR TREE))
               (CONS 'CL (APPEND RIGHT-A-NAMES
                                 (APPEND RIGHT-B-NAMES MPGNAMES?)))))
        '((P) (B-AND PL PR))            ;; The P output.
        '((G) (T-CARRY GL PR GR)))))))))) ;; The G output.
```

Figure 7.11: The PG-ALU Body Generator, TV-ALU-HELP-BODY*

Size	Gate Count	Fanout	Delay
1 bit	126	8	12
2 bits	149	8	14
4 bits	196	8	17
8 bits	297	8	22
16 bits	491	8	26
32 bits	880	8	30
64 bits	1665	8	35
128 bits	3227	8	39

Table 7.3: ALU Characteristics

tion like V-ADDER-BODY, but creates a circuit box that references submodules created by other calls to TV-ALU-HELP* on subtrees of TREE. In other words, the entire boxlist has a recursive structure. This condition is stated formally by the predicate TV-ALU-HELP&.

```
(TV-ALU-HELP& BOXLIST TREE)
 =
(IF (NLISTP TREE)
    (AND (EQUAL (ASSOC (CONS 'TV-ALU-HELP TREE) BOXLIST)
               (TV-ALU-HELP* TREE))
         (T-CELL& (CDR BOXLIST)))
    (AND (EQUAL (ASSOC (CONS 'TV-ALU-HELP TREE) BOXLIST)
               (TV-ALU-HELP* TREE))
         (T-CARRY& (CDR BOXLIST))
         (B-AND& (CDR BOXLIST))
         (V-BUF& (CDR BOXLIST) 8)
         (TV-ALU-HELP& (CDR BOXLIST) (CAR TREE))
         (TV-ALU-HELP& (CDR BOXLIST) (CDR TREE)))))
```

Using an HDL-based approach provides an ability to explicitly control circuit fanout, loading, and delay. Using the functions GEVAL, DEVAL and LEVAL we have computed the gate count, delay, and maximum fanout of our ALU for different word sizes; these are given in Table 7.3. The delay of our ALU with its propagate-generate look-ahead logic grows with the $\log_2$ of the size of the ALU when generated with balanced binary trees.

Using the Boyer-Moore theorem prover on a Sun 3/280, the processing of our ALU proof, including the time necessary to define the primitives and interpreters, takes less than one hour. To generate all of the ALU designs for Table 7.3 only takes a few seconds. A few seconds more are required to check that each ALU is recognized by NEW-ALU& and BOXLIST-OKP.

7.10   Translating our HDL into a Commercial CAD Language

One motivation for defining our HDL was to ease the conversion of verified designs into a form acceptable to commercial CAD systems. Instead of attempting to define our own HDL, we could have attempted to formally define an existing HDL, e.g., VHDL [8]. However, before attempting to formalize a complex language, we wanted to explore the subtleties of defining an HDL without the constraints imposed by an existing language. For example, we did not want to spend time worrying over whether we have the "correct" syntax; any unambiguous syntax will do.

Providing a simple translation into commercial HDLs is important for us as we wish to be able to physically realize our verified circuit descriptions. Because our HDL is simple, we expect that designs specified with our HDL should translate to a commercial HDL in a straightforward manner. However, the accuracy of such a translation is open to question. We do not have meaning functions for any commercial HDL; that is, we do not have a function like HEVAL for any commercial HDL. Therefore, any translation from our HDL to a commercial HDL cannot be verified.[5]

The Boyer-Moore theorem prover is implemented in Common Lisp and manipulates Lisp representations of Boyer-Moore logic terms. The Boyer-Moore theorem prover provides Common Lisp implementations of functions defined in the logic. These Common Lisp procedures are an integral part of the mechanization of the logic, i.e., the soundness of the theorem prover depends upon their correctness. We use the Common Lisp versions of the formal boxlist-creating functions as the basis of our translations. For example, the boxlist-generating function (V-ADDER-BOXLIST N) in the Boyer-Moore logic has a Common Lisp counterpart (*1*V-ADDER-BOXLIST N). Evaluating (*1*V-ADDER-BOXLIST 4) in Common Lisp yields the boxlist below.

```
'(((V-ADDER . 4)
   ((CARRY . 5)
    (A . 4) (A . 3) (A . 2) (A . 1)
    (B . 4) (B . 3) (B . 2) (B . 1))

   ((SUM . 4) (SUM . 3) (SUM . 2) (SUM . 1) (CARRY . 1))

   ((((SUM . 4) (CARRY . 4)) (FULL-ADDER (A . 4) (B . 4) (CARRY . 5)))
```

---

[5]It seems unlikely that we will have a meaning function for a commercial HDL in the near future. Most commercial HDLs that we have investigated are large and complex, and these languages do not have formally defined syntax or semantics. A reasonable approach might be to formally define a subset of a popular commercial HDL. However, even with such a definition the software tools used to create integrated circuit masks from HDL circuit specifications will not be verified.

```
        (((SUM . 3) (CARRY . 3)) (FULL-ADDER (A . 3) (B . 3) (CARRY . 4)))
        (((SUM . 2) (CARRY . 2)) (FULL-ADDER (A . 2) (B . 2) (CARRY . 3)))
        (((SUM . 1) (CARRY . 1)) (FULL-ADDER (A . 1) (B . 1) (CARRY . 2)))))

(FULL-ADDER (A B C) (SUM CARRY)
                (((SUM1 CARRY1) (HALF-ADDER A B))
                 ((SUM CARRY2) (HALF-ADDER SUM1 C))
                 ((CARRY) (B-OR CARRY1 CARRY2))))

(HALF-ADDER (A B) (SUM CARRY)
                (((SUM) (B-XOR A B)) ((CARRY) (B-AND A B)))))
```

To make the translation process concrete, we present the translation of the circuit box above into LSI Logic's NDL. This translation is performed by adding keywords and other bits of syntax to our circuit box expression; translating Lisp lists into comma-separated lists; converting our formal gate names into LSI Logic macrocell names; and translating our CONS-indexed names into a more standard form. The result of converting the boxlist above into NDL is shown in Figure 7.12.

To perform the translation of boxlists into NDL, we have written a translator in Common Lisp. The complete source text for our translator is less than two pages of Lisp code. Since there is no formal model of NDL, we are not able to verify our translator, but we are pleased that it is so short.

7.11 Conclusions

The formalization of even a simple, combinational HDL is more subtle and complex than might first be imagined; however, the usefulness of having circuits represented as data cannot be over-emphasized. Our HDL enjoys the benefits of being purpose-built for concisely and directly expressing hardware circuits, but incurs the costs associated with having to prove the correctness of circuit operation through an interpreter.

We believe that it is possible to formalize features of commercial CAD languages. If this is possible, then we may be able to adapt existing verification systems to support commercial CAD systems. We do not expect commercial CAD system suppliers to convert to formal systems (e.g., HOL, Boyer-Moore, NuPRL) without these systems being able to support the kinds of activities commercial customers now enjoy.

The use of arbitrary heuristics is possible for provably correct circuit generating functions. For instance, consider a heuristic that chooses between two correct implementations. The heuristic may perform any amount of analysis, but no matter what choice is returned the resulting circuit generated will still be correct. We employed a simple algebraic choice heuristic in the verification

```
COMPILE;
DIRECTORY MASTER;

MODULE V-ADDER_4;
INPUTS CARRY_5,A_4,A_3,A_2,A_1,B_4,B_3,B_2,B_1;
OUTPUTS SUM_4,SUM_3,SUM_2,SUM_1,CARRY_1;
LEVEL FUNCTION;
DEFINE
G_0(SUM_4,CARRY_4) = FULL-ADDER(A_4,B_4,CARRY_5);
G_1(SUM_3,CARRY_3) = FULL-ADDER(A_3,B_3,CARRY_4);
G_2(SUM_2,CARRY_2) = FULL-ADDER(A_2,B_2,CARRY_3);
G_3(SUM_1,CARRY_1) = FULL-ADDER(A_1,B_1,CARRY_2);
END MODULE;

MODULE FULL-ADDER;
INPUTS A,B,C;
OUTPUTS SUM,CARRY;
LEVEL FUNCTION;
DEFINE
G_0(SUM1,CARRY1) = HALF-ADDER(A,B);
G_1(SUM,CARRY2) = HALF-ADDER(SUM1,C);
G_2(CARRY) = OR2(CARRY1,CARRY2);
END MODULE;

MODULE HALF-ADDER;
INPUTS A,B;
OUTPUTS SUM,CARRY;
LEVEL FUNCTION;
DEFINE
G_0(SUM) = EO(A,B);
G_1(CARRY) = AN2(A,B);
END MODULE;

END COMPILE;
END;
```

Figure 7.12: NDL Translation of a Four-bit Adder

of our ALU generator; this heuristic helps control the fanout of the control lines. It is easy to imagine much more sophisticated heuristics.

Modeling circuits as data in a formal logic has important benefits: well-formed circuits can be formally specified, programs can be used to manipulate circuits, circuits can be verified, and the verification of circuit manipulating tools (e.g., a logic minimizer) can be accomplished. That is, it is possible to build a system where verified circuits are created with the assistance of verified tools. For instance, a verified synthesis program could be constructed, thus eliminating the need for after-the-fact proofs. Our ALU circuit generator is an instance of a verified synthesis program.

We believe the formalization of as much of the design space as possible is critical to the production of higher quality hardware. We are presently extending our HDL to include registers, tri-state and open-collector logic, and other features germane to digital hardware design. It is important that we continue to extend our formal models to include any notion of hardware specification that is important to correctly build computing hardware.

Even though the notion of a formula manual is many years off, we believe formal methods will play an increasingly important role in the construction of predictable computer hardware. The ability to provide multiple interpretations for circuits expressed with our HDL is a step toward providing greater formal modeling coverage of physical properties important to correct hardware device operation.

7.12 References

[1] Bishop C. Brock and Warren A. Hunt, Jr. The Formalization of a Simple HDL. In *Proceedings of the IFIP TC10/WG10.2/WG10.5 Workshop on Applied Formal Methods for Correct VLSI Design.* Elsevier Science Publishers, 1990.

[2] R. S. Boyer and J S. Moore. *A Computational Logic Handbook.* Academic Press, Boston, 1988.

[3] Geoffrey M. Brown and Miriam E. Lesser. From Programs to Transistors: Verifying Hardware Synthesis Tools. In *Workshop on Hardware Specification, Verification and Synthesis: Mathematical Aspects.*, volume 408 of *Lecture Notes in Computer Science*, pages 128–150. Springer Verlag, 1989.

[4] Randal E. Bryant. Verification of Synchronous Circuits by Symbolic Logic Simulation. In *Hardware Specification, Verification and Synthesis: Mathematical Aspects.*, volume 408 of *Lecture Notes in Computer Science*, pages 14–24. Springer Verlag, 1989.

[5] Avra Cohn. Correctness Properties of the VIPER Block Model: The Second Level. In *Current Trends in Hardware Verification and Automatic Theorem Proving*, pages 1–91. Springer-Verlag, New York, 1989.

[6] W. J. Cullyer and C. H. Pygott. Application of Formal Methods to the VIPER Microprocessor. *IEE Proceedings*, 134, Pt. E(3):133–141, May 1987.

[7] Diederik Verkest, Luc Claesen, and Hugo De Man. The Use of the Boyer-Moore Theorem Prover for Correctness Proofs of Parameterized Hardware Modules. In *Proceedings of the IFIP TC10/WG10.2/WG10.5 Workshop on Applied Formal Methods for Correct VLSI Design*. Elsevier Science Publishers, 1990.

[8] Editors: Larry Saunders and Ronald Waxman. *IEEE Standard VHDL Language Reference Manual*. The Institute of Electrical and Electronics Engineers, 1988.

[9] Joesph A. Goguen. OBJ as a Theorem Prover with Applications to Hardware Verification. In *Current Trends in Hardware Verification and Automatic Theorem Proving*, pages 218–267. Springer-Verlag, New York, 1989.

[10] M. Gordon. HOL: A Proof Generating System for Higher-Order Logic. Technical Report 103, University of Cambridge, Computer Laboratory, 1987.

[11] M.J.C. Gordon. Why Higher-Order Logic is a Good Formalism for Specifying and Verifying Hardware. Technical Report 77, University of Cambridge, Computer Laboratory, September 1985.

[12] Graham Birtwistle, Brian Graham, Todd Simpson, Konrad Slind, Mark Williams, and Simon Williams. Verifying an SECD chip in HOL. In *Proceedings of the IFIP TC10/WG10.2/WG10.5 Workshop on Applied Formal Methods for Correct VLSI Design*. Elsevier Science Publishers, 1990.

[13] Guy L. Steele, Jr. *Common LISP: The Language*. Digital Press, 1984.

[14] Steven D. Johnson. Manipulating Logical Organization with System Factorizations. In *Hardware Specification, Verification and Synthesis: Mathematical Aspects.*, volume 408 of *Lecture Notes in Computer Science*, pages 259–280. Springer Verlag, 1989.

[15] Mike Gordon, Albert Camilleri, and Tom Melham. Hardware Verification Using Higher-Order Logic. Technical Report 91, University of Cambridge Computer Laboratory, September 1986.

[16] Bill Pace and Mark Saaltink. Formal Verification in m-EVES. In *Current Trends in Hardware Verification and Automatic Theorem Proving*, pages 268–302. Springer-Verlag, New York, 1989.

[17] Mary Sheeran. μFP - An Algebraic VLSI Design Language. Technical Report PRG-39, Oxford Programming Research Group, September 1984.

[18] Simon Finn, Michael P. Fourman, Michael Francis, and Robert Harris. Formal System Design - Interactive Synthesis based on Computer-Assisted Formal Reasoning. In *Proceedings of the IFIP TC10/WG10.2/WG10.5 Workshop on Applied Formal Methods for Correct VLSI Design*. Elsevier Science Publishers, 1990.

[19] Warren A. Hunt, Jr. FM8501: A Verified Microprocessor. Technical Report ICSCA-CMP-47, University of Texas at Austin, 1985.

[20] Warren A. Hunt, Jr. and Bishop C. Brock. The Verification of a Bit-Slice ALU. In *Workshop on Hardware Specification, Verification and Synthesis: Mathematical Aspects.*, Lecture Notes in Computer Science, pages 281–305. Springer Verlag, 1989.